SCARCE MIGRANT BIRDS IN BRITAIN AND IRELAND

To

W. G. Jarvis,

James Ferguson-Lees and Ken Williamson,

in gratitude for their tuition,

encouragement and example

By the same author (editor):

THE NATURAL HISTORY OF CAPE CLEAR ISLAND

Scarce migrant birds in Britain and Ireland

J. T. R. SHARROCK

WITH LINE DRAWINGS BY

P. J. GRANT

T. & A. D. POYSER
Berkhamsted

First published 1974 by T & AD Poyser Ltd
Print-on-demand and digital editions published 2010 by T & AD Poyser,
an imprint of A&C Black Publishers Ltd, 36 Soho Square, London W1D 3QY

www.acblack.com

ISBN (print) 978-1-4081-3738-3
ISBN (epub) 978-1-4081-3736-9
ISBN (e-pdf) 978-1-4081-3737-6

A CIP catalogue record for this book is available from the British Library

Printed and bound in Great Britain

This book is largely based on material which first appeared in the journal *British Birds*. The reference for that material is as follows:
J. T. R. Sharrock 1969–73: Scarce migrants in Britain and Ireland during 1958-67. *British Birds*, 62: 169-189,300-315; 63: 6-23, 313-324; 64: 93-113, 302-309; 65: 187-202,381-392; 66: 46-64, 517-525.

Contents

List of species

List of Plates

(between pages 16 and 17)

The photographs by R. H. Dennis are of birds caught for ringing at Fair Isle Bird Observatory

LINE DRAWINGS

The line drawings by P. J. Grant show the species named in the section heading, with the following exceptions: page 92, Pectoral Sandpiper (left) and Semipalmated Sandpiper (right); page 108, Myrtle Warbler; page 116, Arctic Warbler (top) and Greenish Warbler (bottom). The drawings of Sabine's Gull (page 103) and White-winged Black Tern (page 131) show the distinctive immature plumages

Acknowledgements

Very special thanks are due to I. J. Ferguson-Lees for his constant encouragement and for help in innumerable ways; to my wife for her tolerance of my extraordinary working hours, for her frequent advice and for translating the foreign papers for me; to P. A. D. Hollom and William Collins Sons & Co. Ltd. for permission to include European distribution maps from the *Field Guide*; to Miss Karen Rayner for expertly redrawing all of the histograms in Chapters 5–9; and to P. F. Bonham for valuable assistance, particularly with Chapter 10.

The series of papers which first appeared in the journal *British Birds* is now published in book form with the permission of the editors and of Macmillan Journals Ltd., to whom I express my thanks.

I am most grateful to R. H. Dennis, W. Gatter and R. E. Scott who each supplied me with typescripts of their work in advance of publication.

The data were punched and verified by Dynamic Data Ltd and I am grateful to the Managing Director, Mr Dann, for arranging special terms and to Antony Witherby for arranging that *British Birds* should cover the cost. The useful idea of compiling a map of observer distribution (Fig. 1), to give substance to a textual comment, came from Stanley Cramp. The standard map outlines for the county and regional distributions were drawn by Robert Gillmor.

I extracted the data mainly from county and regional reports, and I believe that observers and readers of this book should remember the enormous amount of work which local recorders and report editors put into the preparation of those invaluable documents in their free time. A number of editors allowed me to see typescripts of their 1967 reports in advance of publication and many more were helpful in promptly answering my queries. Over 100 unpublished records were received directly from observers and these data were incorporated here after they had been passed to the relevant county recorder and verified. The following list acknowledges a lot of essential help from many people, to all of whom I express my thanks: D. G. Andrew, H. E. Axell, Mrs R. G. Barnes, J. H. Barrett, Rev J. E. Beckerlegge, D. G. Bell, T. Hedley Bell, Dr W. R. P. Bourne, W. F. A. Buck, E. A. Chapman, B. M. Church, F. R. Clafton, C. J. Coe, J. Cudworth, P. Davis, R. H. Dennis, A. Dobbs, J. W. Donovan, G. M. S. Easy, T. Ennis, G. Evans, J. K. Fenton, Dr J. E. C. Flux, J. C. Follett, G.

Harris, T. Hooker, Mrs J. Hope Jones, P. Hope Jones, M. T. Horwood, R. Hudson, Dr D. Jenkins, F. H. Jones, M. Jones, Brigadier S. T. M. Kent, E. D. Kerruish, Mrs K. M. Kirton, Mrs R. F. Levy, W. G. Lewis, J. Lord, A. T. Macmillan, The Earl of Mansfield, C. J. Mead, B. S. Milne, J. Mitchell, J. L. Moore, Mrs J. R. Naish, Dr I. C. T. Nisbet, T. R. H. Owen, Miss E. M. Palmer, Dr J. D. Parrack, J. L. F. Parslow, A. G. Parsons, R. J. Partridge, W. H. Payn, Lt-Col H. R. Perkins, K. Preston, Dr N. O. Preuss, Miss H. M. Quick, J. Rabøl, R. A. Richardson, C. S. Robbins, T. Ruck, Major R. F. Ruttledge, R. E. Scott, M. J. Seago, M. Shrubb, F. R. Smith, Mrs I. Smith, R. Spencer, D. M. Stark, R. Stokoe, D. D. B. Summers, C. M. Swaine, J. H. Taverner, A. D. Townsend, R. J. Tulloch, A. J. Vickery, K. G. Walker, C. Waller, A. J. Wallis, Dr A. Watson, G. Webber, Mr and Mrs J. K. Weston, C. N. Whipple, A. A. K. Whitehouse, and G. A. Williams.

Introduction

The network of bird observatories round our coasts and the fieldwork of many individual birdwatchers have resulted in a vast accumulation of information on the migrants which visit Britain and Ireland. The analysis of records from individual stations by local enthusiasts, and the derivation of the important underlying principles by such well-known names as Peter Davis, Dr I. C. T. Nisbet and Kenneth Williamson, brought a new understanding to the phenomena we observe. Information from radar work by Dr David Lack and other workers in the late 1950s and early 1960s necessitated the reconciling of the resulting new theories with those of 'down-wind drift', propounded especially by Kenneth Williamson from his work as Warden of Fair Isle Bird Observatory and as Migration Research Officer of the British Trust for Ornithology; and a further advance was the theory of 'reverse migration', especially fostered here by Dr I. C. T. Nisbet. Since those halcyon days, data have continued to accumulate.

The common migrants, such as Willow Warbler *Phylloscopus trochilus* and Wheatear *Oenanthe oenanthe*, occur in such large numbers all around our coasts, as well as inland, that we can collect data only from selected sites, such as bird observatories. Thanks to the journal *British Birds* and their Rarities Committee formed in 1958, we now have annual lists of the major rarities, all checked and verified by 'The Ten Rare Men', as the ten members of that committee are affectionately (if sarcastically) called. Between these two groups of birds, the very common and the very rare, lie those which are here named 'the scarce migrants'—those which occur annually, or nearly so, in numbers ranging from a handful to a hundred or more. Nearly every record of these scarce migrants is published in the excellent series of annual county and regional bird reports which covers the whole of Britain and Ireland, and they supply ideal data for detailed analysis. This book, based on a series of papers in the journal *British Birds* (Sharrock 1969–73), takes 24 of these scarce migrants (plus groups of American waders and passerines) and analyses their records in Britain and Ireland during one ten year period (1958–67: referred to throughout this book as 'the ten years'). The records of the rarer species were extracted from the reports of the Rarities Committee and the *Irish Bird Report*, while those of the commoner species* were extracted from the county and regional reports. The only practicable method was to collate these latter records as published in regional reports without attempting to assess their validity, on the assumption that they had already been adequately vetted by the county recorders. In many cases, local reports only summarised records, but the editors concerned were extremely helpful in supplying fuller details. Also, in an attempt to fill in gaps, an appeal was made (*Brit. Birds*, 61 : 470–471) for unpublished observations. This resulted in the receipt of over 100 further records which, as well as being included in these analyses, were passed on to the relevant county recorders. A few additional records have come to light since these analyses were carried out in 1969–73, but would not change the general patterns or alter the conclusions given here.

To facilitate analysis, the data for each record were coded and the analyses carried out by means of Hollerith punched cards, using 18 columns. The data coded for every record were as follows:

1–2	Species code number	13	Number of immatures
3–4	Year	14	Number of males
5–7	County code number	15	Number of females
8–9	Regional code number	16–17	Coded date
10–11	Number of individuals	18	Additional data (e.g. seen from boat, breeding, sub-species)
12	Number of adults		

In these analyses the basic unit used for displaying distribution within Britain and Ireland is the county. It was not practicable to pinpoint each record: even though a majority were at bird observatories, the large number of obscure gravel pits and other local names (often not even shown on one-inch Ordnance Survey maps) made precise location of some 7,000 records impossible. Thus the use of a grid system (e.g. 100 km squares) was impracticable. In preparing the distribution maps, I have used a wide range of symbol sizes (usually 14) to give an impression of the relative abundance in different areas. Arbitrary decisions had sometimes to be made when localities were on county boundaries. These were few, the only ones occurring more than once being Wisbech sewage farm (Norfolk/Lincolnshire) and Rye Meads (Hertfordshire/Essex), both of which were always treated as in the first-

* Rough-legged Buzzard, Temminck's Stint, Pectoral Sandpiper, Long-tailed Skua, Mediterranean Gull, Sabine's Gull, Hoopoe, Golden Oriole, Bluethroat, Melodious Warbler, Icterine Warbler, Barred Warbler, Yellow-browed Warbler, Red-breasted Flycatcher and Ortolan Bunting.

named county. Rutland was incorporated with Leicestershire and the Isle of Wight with Hampshire, but the Isles of Scilly were treated as a county in their own right and separate from Cornwall. While the county was the geographical unit generally employed, it was sometimes convenient to use larger areas and, for this purpose, Britain and Ireland were divided into twelve regions (shown by heavy lines in Fig. 1) as follows:

1. *South-east England:* Kent, London, Surrey, Sussex, Hampshire
2. *South-west England:* Dorset, Wiltshire, Somerset, Devon, Cornwall, Scilly
3. *East Anglia:* Essex, Suffolk, Norfolk, Cambridgeshire
4. *Eastern England:* Lincolnshire, Yorkshire
5. *North-east England:* Durham, Northumberland
6. *North-west England:* Cheshire, Lancashire, Westmorland, Cumberland, Isle of Man
7. *Midlands:* other English counties
8. *Wales*
9. *South of Ireland:* Munster and Leinster } Not corresponding to the Republic of Ireland and Northern Ireland
10. *North of Ireland:* Ulster and Connacht }
11. *Northern Scotland:* Scotland north from Angus, Perthshire and Argyll
12. *Southern Scotland:* Scotland south of Angus, Perthshire and Argyll

Where a species is relatively common, records are sometimes best dealt with as 'bird-days', but this information was often not available. In any event, it was considered more interesting to know the actual number of individuals occurring. Often the only information recorded was the arrival date of each bird. It was therefore easier, as well as preferable, to assess the number of individuals from 'bird-day' data than to try to compose 'bird-day' data from records of individuals. Where the records were summarised in a county report, the local editor was asked for his assessment of the number of individuals concerned. For instance, ringing or plumage variation might show that a series of records such as 5, 3, 2, 6 and 1 on successive days referred to any number from six to 17 individuals. Such decisions were always left to the county editor or observatory warden who had access to all the relevant data and knew the local conditions.

The published details were always extracted with no added assumptions. For example, a male Golden Oriole was not assumed to be an adult unless this was specifically stated. Where two or more reports

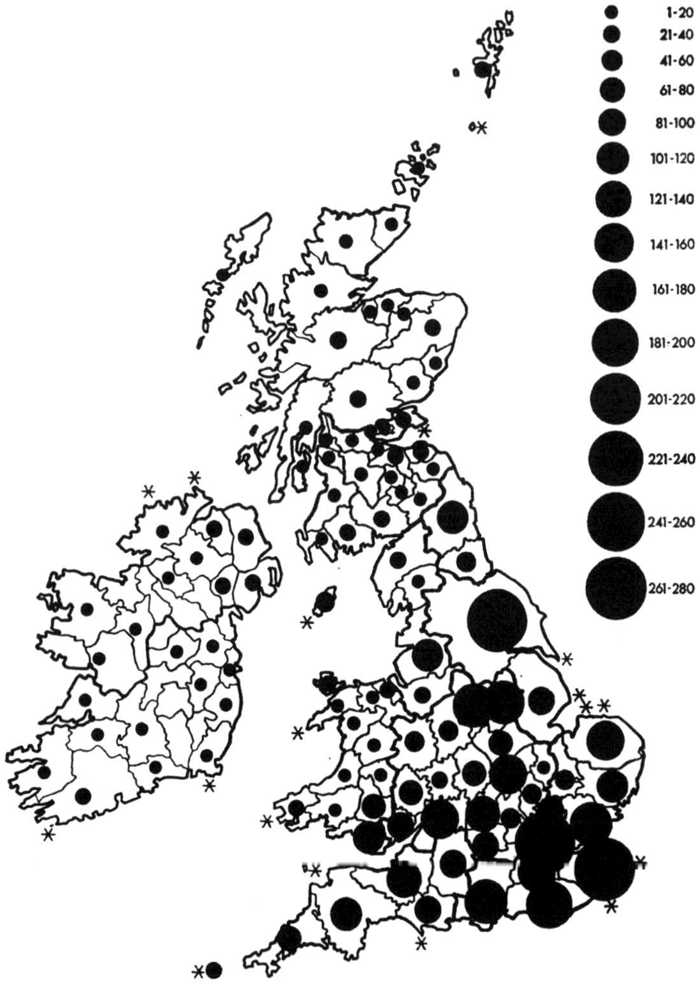

Fig. 1. Distribution by counties of observers in Britain and Ireland. The dots show the actual or estimated number of contributors to the 1967 (sometimes 1966) county bird report except in Scotland, Ireland and a few other areas where estimates were made by the regional editors (see text). The stars show bird observatories which were in operation for a significant part of the ten years and which had abnormally good coverage either through a resident warden or through drawing largely on observers from outside the county. The twelve regional divisions used are outlined more heavily than the county boundaries

overlapped and covered the same area, a record published in only one report was included in the analysis unless the other(s) specifically rejected it; this occurred mainly in the earlier years, when liaison between neighbouring editors was less good than it is now.

It should be pointed out that the totals may not always correspond in the various analyses of a species. In a number of cases the date of an occurrence is not known beyond, for example, 'mid-summer'; even the exact year may not be determined, for example in belated records from reliable non-ornithologists of such species as Hoopoe or Golden Oriole; and in some instances the county is not known, where a report covers two or more counties and the present editor has not heard of a locality published in an earlier report. Records of these kinds are very few, but in such cases a record which can be included in a histogram showing seasonal or yearly distribution cannot be shown in a map of geographical distribution, or vice versa.

The distribution throughout the year is always displayed in seven-day periods to eliminate the effect of weekend-bias (Sharrock 1966). The data for the ninth period (26th February–4th March) are always adjusted to compensate for leap-years ($\times \frac{70}{72}$) and those for the 52nd period (24th–31st December) to compensate for this being an eight-day period ($\times \frac{7}{8}$).

Readers should bear in mind two factors which greatly affect these analyses. First, a great increase in interest in ornithology in recent years is shown by, for instance, the growth in membership of the British Trust for Ornithology from 2,667 in 1958 to 4,456 in 1967.* The yearly totals of a species should, therefore, be judged with this in mind. Secondly, the distribution of observers is far from even, so that there are, for example, more active recorders in an average south-east English county than in the whole of Ireland. The geographical distribution of migrant records must partly reflect this uneven cover. At the same time, observatories and ringing stations contribute a high proportion of the passerine records and so the establishment or closure of one of these centres can alter the whole pattern within a county. To give some idea of this uneven observer distribution, Fig. 1 plots the number of people in each county who contributed to the 1967 (sometimes 1966) county report. (In the cases of the Republic of Ireland, Scotland, Northern Ireland, parts of Wales and a few English counties, where recording is not on a county basis, these figures were estimated for me by the regional editors.) These data must not be regarded as completely accurate, since each editor undoubtedly has differing standards when

* B.T.O. membership exceeded 6,700 in 1973.

defining a 'contributor', but the aim is only to show the bias due to uneven observer distribution. Bird observatories which had abnormally good coverage, either through a resident warden or because they drew largely on observers from outside the county in which they are situated, are shown by stars. I am grateful to J. N. Dymond for advice regarding the selection of such observatories to be included for this ten-year period.

It should be pointed out that detailed analysis of an exceptional movement in any one year is not one of the aims of these analyses. Items of such immediate interest usually warrant papers in their own right (e.g. Scott 1968) and, while these analyses will refer to such papers, they are not intended to take their place.

The vernacular names, scientific names and (where relevant) the sequence of species follow Hudson (1971). In order to facilitate comparisons between species, the chapters in this book deal in turn with natural groupings of species having similar breeding distributions or other features in common.

Although the original material was written for the serious student of migration, I hope that this book will now reach a wider audience, for there must be many who enjoy watching birds and wish to know more about them, who never read all the scientific journals. I have, therefore, briefly described the birds dealt with, in a few introductory remarks to each chapter. The reader who wishes further information on identification should consult one of the several field guides now available. I particularly commend *A Field Guide to the Birds of Britain and Europe* by R. T. Peterson, Guy Mountfort and P. A. D. Hollom (Collins 1954, revised 1966), from which the European distribution maps in this book were taken.

I Rough-legged Buzzard *Buteo lagopus*, hovering (Ján Švehlik)
Temminck's Stint *Calidris temminckii*, on nest (J. B. & S. Bottomley)

II Pectoral Sandpiper *Calidris melanotos* (J. B. & S. Bottomley)
Long-tailed Skua *Stercorarius longicaudus* (Eric Hosking)

III Mediterranean Gull *Larus melanocephalus*, second year(J. B. & S. Bottomley)
Sabine's Gull *Larus sabini*, adult off western Ireland (J. W. Enticott)

IV White-winged Black Tern *Chlidonias leucopterus* (K. Atkin)
Gull-billed Tern *Gelochelidon nilotica*, immature (K. Atkin)

V Hoopoe *Upupa epops*, at nest (Eric Hosking); and Golden Oriole *Oriolus oriolus*, male at nest (M. D. England)

VI Bluethroat *Luscinia svecica*, red-spotted (M. D. England)
Aquatic Warbler *Acrocephalus paludicola* and Sedge Warbler *A. schoenobaenus* (D. J. Steventon)

VII Melodious Warbler *Hippolais polyglotta*, at nest (M. D. England)
Icterine Warbler *Hippolais icterina*, at nest (Eric Hosking)

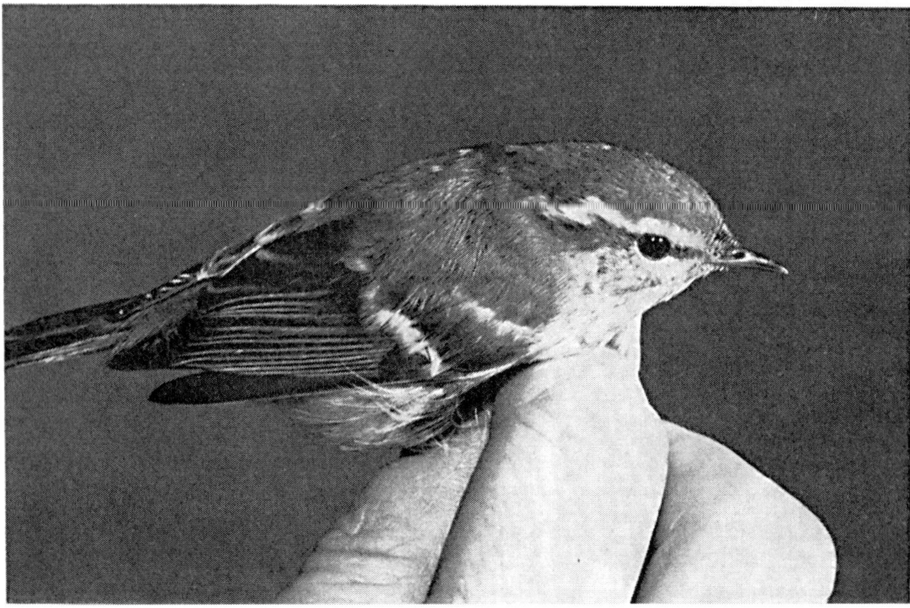

VIII Barred Warbler *Sylvia nisoria*, immature (K. Atkin)
Yellow-browed Warbler *Phylloscopus inornatus* (Eric Hosking)

IX Greenish Warbler *Phylloscopus trochiloides* (R. H. Dennis)
Arctic Warbler *Phylloscopus borealis* (R. H. Dennis)

X Red-breasted Flycatcher *Ficedula parva* (Dr. C. M. Perrins)
Woodchat Shrike *Lanius senator*, immature (Malcolm Wright)

XI Richard's Pipit *Anthus novaeseelandiae* (R. H. Dennis)
Tawny Pipit *Anthus campestris* (P. O. Swanberg)

XII Scarlet Rosefinch *Carpodacus erythrinus* (R. H. Dennis)
Ortolan Bunting *Emberiza hortulana* (R. H. Dennis)

CHAPTER 1

Hoopoe, Golden Oriole and Tawny Pipit

These three species all have breeding distributions covering almost the whole of Continental Europe, except for Scandinavia (Figs. 5, 11 and 15).

Hoopoes are striking black-and-white-and-orange birds, slightly larger than a Mistle Thrush *Turdus viscivorus*, with a large erectile crest and rounded wings. They breed in orchards and woodland with open glades but on passage are usually seen feeding on short turf. Their distinctive appearance and not infrequent occurrence on lawns leads to their being reported quite often by members of the general public, even those who normally seldom take note of birds.

Although male Golden Orioles are equally striking—brilliant yellow, with black on the wings and tail—these birds, slightly larger than Song Thrushes *T. philomelos*, are shyer than Hoopoes and are often detected only by their fluty calls emanating from thick woodland. The females and young males are usually less bright, with greenish-yellow plumage. On passage, when newly-arrived in coastal areas, they may frequent open ground, the males' plumage closely matching the flowers of gorse bushes in such areas. Although unmistakable when seen well, tyros sometimes misidentify poorly-seen Green Woodpeckers *Picus viridis* as this species.

Tawny Pipits, by comparison with the other two species considered here, are dull birds. In appearance rather like short-tailed wagtails, the adults are distinctive (as pipits go), with mainly uniform sandy plumage, apart from dark centres to the median wing-coverts, forming a darkish wing-bar. Young birds, however, are heavily streaked and superficially resemble Richard's Pipits (see Chapter 8), though with a different call and more horizontal stance (Grant 1972). On passage, they are most often seen feeding on short turf near the sea.

Hoopoe

Upupa epops

A total of 1,245 Hoopoes was recorded in Britain and Ireland during 1958–67, an average of 125 per year. The vast increase in watching (and recording) which has taken place in the last 20 years makes it difficult to compare numbers with those recorded earlier, but it is remarkable that 31 Hoopoes in 1948 and 30–34 in 1950 were in both cases sufficient to stimulate special summaries in the journal *British Birds* (Editors 1949, 1951).

Whilst there were records in every month of the year, occurrences were very markedly concentrated in the periods from the end of March to mid-May and from August to September, with the peaks in mid-April and early September (Fig. 2). The spring passage was very much stronger than the autumn one. Arbitrarily separating the spring and autumn passages between the 26th and 27th seven-day periods (up to 1st July and from 2nd July onwards), 77% of the occurrences fell within the first half of the year and 23% within the second half. This is a greater difference between the seasons than there was in the hundred years 1839–1938, when 67% were recorded in spring and 33% in autumn (derived from Glegg 1942).

The volume of spring passage has varied widely in the ten years, from 34 individuals in 1961 to 154 in 1965 (Fig. 3). The spring records are somewhat suggestive of a cyclical pattern, a peak in 1958 declining to a trough in 1961, steadily rising to a peak in 1965 and then falling again. This may be fortuitous, however, for arrivals have been linked with those of other vagrants appearing either with southerly winds and overcast associated with fronts moving into the English Channel or in clear anticyclonic conditions (Williamson 1961, Davis 1964), the occurrence of these weather patterns perhaps largely determining the number of records of Hoopoes.

Except in 1958, autumn numbers have varied less (Fig. 3), showing a small but steady annual increase in records. This may merely reflect the increase in observers over the ten years, but the observer-increase probably affects Hoopoe numbers less than almost any other species because of the larger proportion recognised and reported by the general public. The numbers in autumn 1958 were so aberrant that it is worth looking at them in more detail. In fact, they were almost entirely the result of an influx which reached a peak during 3rd–9th September (a third of the records falling within this seven-day period) and which was largely restricted to the coastal counties from the Isles of Scilly to Sussex (60% of that autumn's records being in these six counties). The exceptional nature of this autumn influx of Hoopoes appears not to have been fully appreciated at the time (no doubt partly because relatively few were seen at the bird observatories). The period brought large numbers of migrants and many vagrants to both the east coast

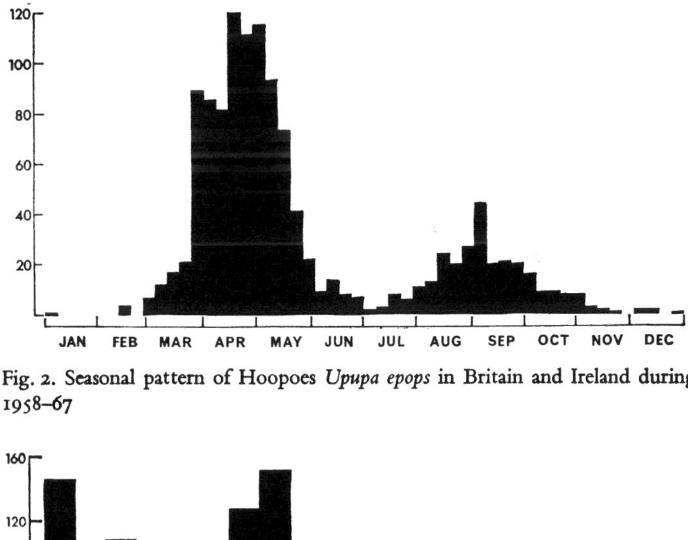

Fig. 2. Seasonal pattern of Hoopoes *Upupa epops* in Britain and Ireland during 1958–67

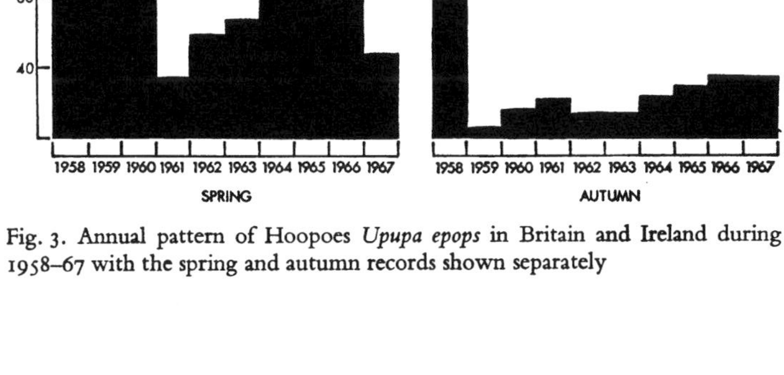

Fig. 3. Annual pattern of Hoopoes *Upupa epops* in Britain and Ireland during 1958–67 with the spring and autumn records shown separately

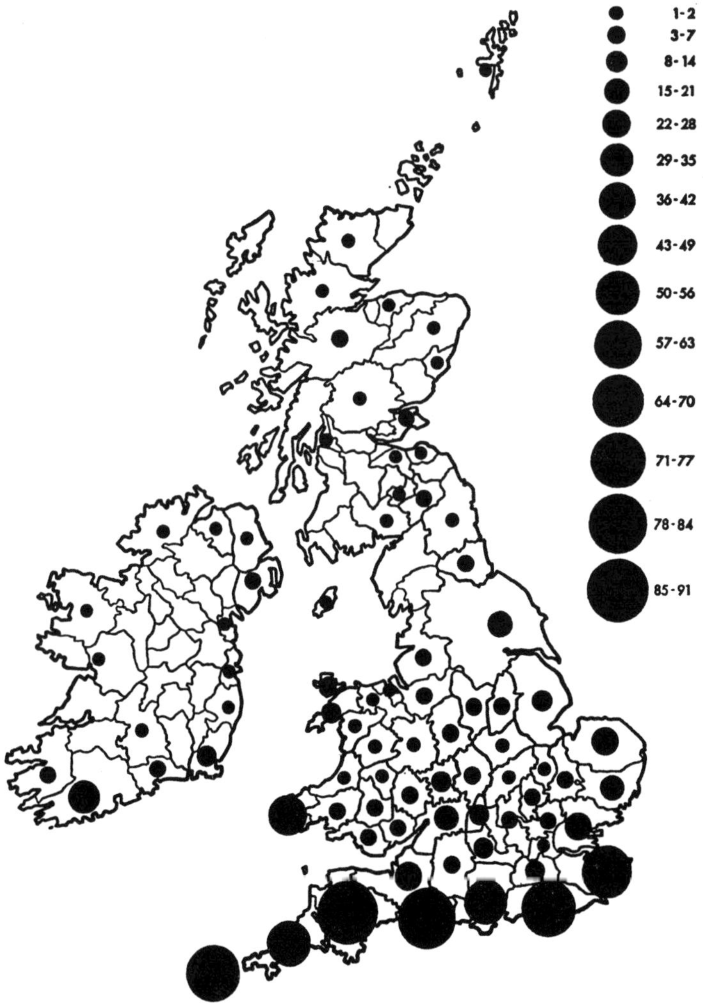

Fig. 4. Distribution by counties of spring Hoopoes *Upupa epops* in Britain and Ireland during 1958–67

and the Irish Sea area (Williamson 1958, 1959), but it is significant that only 14% of the Hoopoes were on the English and Scottish east coasts.

In spring, over the ten years, Hoopoes occurred mainly in the seven counties of the English south coast and in Pembrokeshire and Co.

Cork (Fig. 4). Their frequency in Pembrokeshire was somewhat surprising, as one might have expected that county, shielded from southerly arrivals by the south-west peninsula, to receive relatively few. Whilst these nine counties provided the majority of the spring Hoopoes, reports were very widespread, including every Midland county. Indeed, only two English counties (Cumberland and Westmorland) and no Welsh counties were without a record. The paucity of reports in Ireland (apart from the four southern counties from Cos. Kerry to Wexford) and Scotland was no doubt due in part to the relative lack of observers, though most of the coastal Irish counties and the eastern Scottish counties did have at least one record.

The preponderance of south coast records is not surprising, in view of the breeding distribution of this species which is mainly a summer visitor to Europe (Fig. 5). It should be added, however, that the Hoopoe has decreased markedly in Europe in the last hundred years; this was attributed by Fjerdingstad (1939) to competition for nest sites with the increasing numbers of Starlings *Sturnus vulgaris*, but is more generally considered to be due to the climatic changes which have resulted in cooler, wetter summers in western Europe. In any event, the species is now largely absent as a breeding bird from the Low Countries and Scandinavia.

In Britain and Ireland, taking the ten years as a whole, the earliest arrivals were in the south-west in late February and early March (Fig. 6). By mid-March a few reached south-east England and East Anglia, but the majority was still found in the south-west. From the

Fig. 5. European distribution of Hoopoes *Upupa epops* with the breeding range shown in black and the northern limit of the wintering area marked by a dotted line (reproduced, by permission, from the 1966 edition of the *Field Guide*)

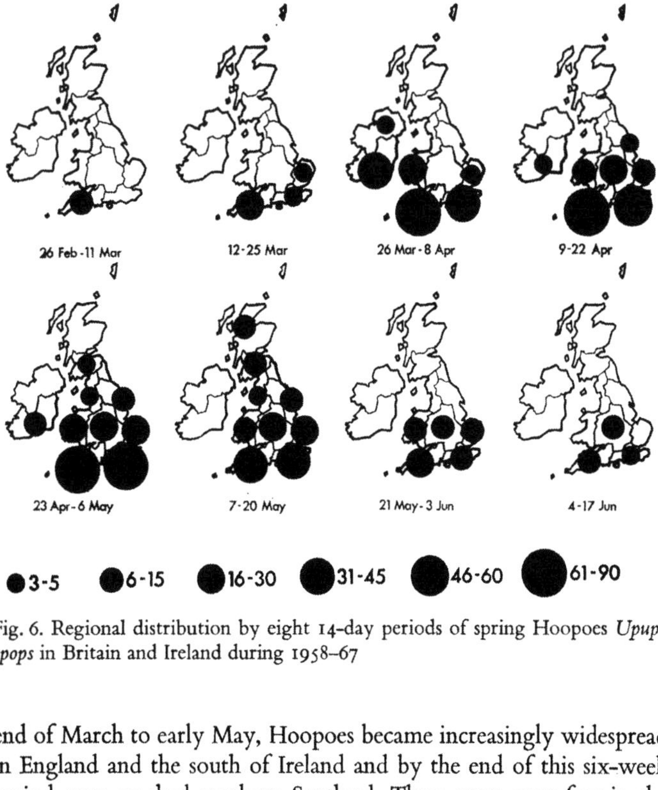

Fig. 6. Regional distribution by eight 14-day periods of spring Hoopoes *Upupa epops* in Britain and Ireland during 1958–67

end of March to early May, Hoopoes became increasingly widespread in England and the south of Ireland and by the end of this six-week period even reached southern Scotland. There were very few in the Midlands until the second fortnight of this six-week period, even though they were widespread in the coastal counties bordering on the Midlands in the first fortnight. This suggests that those seen in the Midlands from the second week of April onwards were largely ones which had arrived previously in the coastal counties and had wandered within Britain. (The alternative, that Hoopoes arriving later in the season penetrated Britain more deeply on arrival than those occurring earlier, is a less logical explanation.) By the second and third weeks of May, numbers had begun to decline, but some reached northern Scotland at this time. At the end of May and in early June, records were virtually confined to southern England, especially the south coast counties and the Midlands. It is likely that these were not new arrivals, but consisted largely of non-breeders summering in Britain.

The spring influxes of Hoopoes in Britain may therefore be broken down into three phases: (1) initial arrivals in the south-west; (2) arrivals further along the English south coast and in the south-east, and continuing arrivals in the south-west; and (3) arrivals along the whole of the east coast, increasingly more northerly as time progresses. It seems

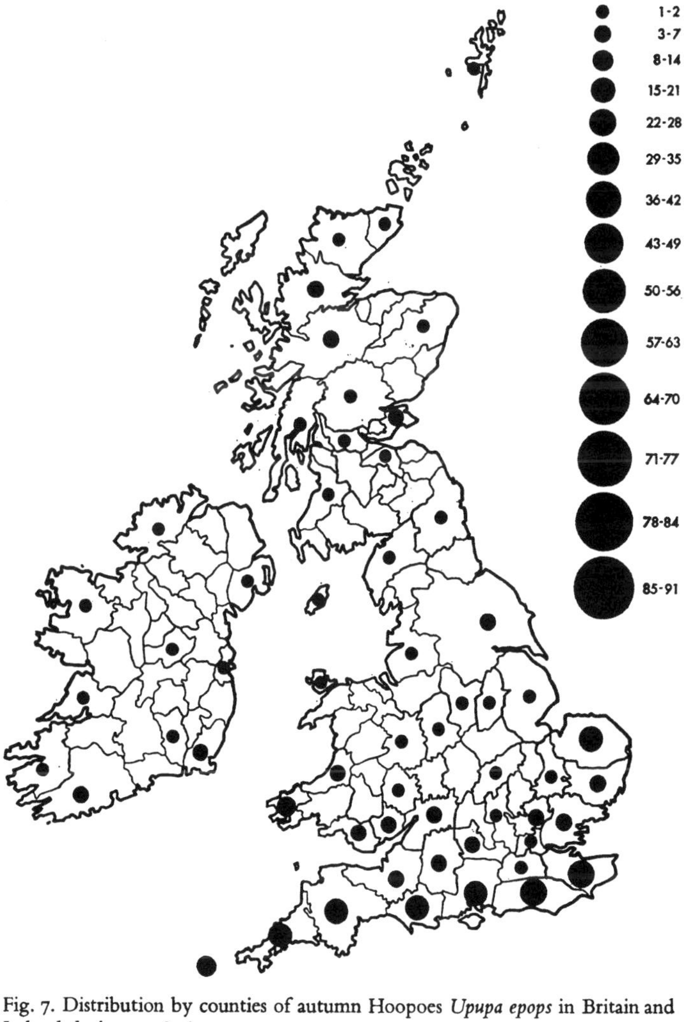

Fig. 7. Distribution by counties of autumn Hoopoes *Upupa epops* in Britain and Ireland during 1958–67

probable that those in March are mainly Iberian birds reaching south-west England, the south of Ireland and south Wales by displacement on a northerly (or sometimes even NNE) track and that the later arrivals in April and early May originate from progressively more northerly and easterly populations, until those in Scotland in May are arriving on an almost north-westerly track.

Relatively few Hoopoes ever remain in Britain to breed. According to Parslow (1973) there has been little change in the average frequency of about one record per decade since the 1830s, with seven in 1895–1904 and four in the 1950s. *The Handbook* notes that the species has nested in every south coast English county, perhaps most often in Hampshire. In the ten years there was only one authenticated breeding record, in Kent in 1959, but a pair summered and breeding was suspected in Bedfordshire in 1964.

It has already been noted that more than three times as many Hoopoes were recorded in spring as in autumn in the ten years, and that over a quarter of the autumn total was seen in a single year (1958).

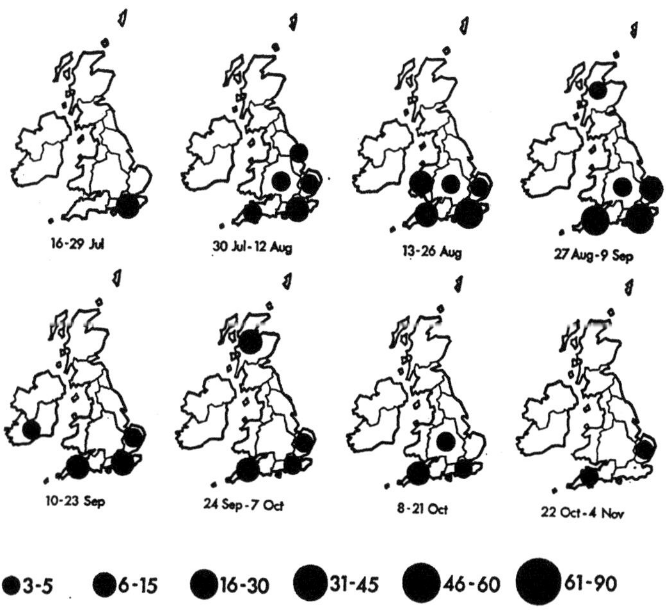

Fig. 8. Regional distribution by eight 14-day periods of autumn Hoopoes *Upupa epops* in Britain and Ireland during 1958–67

The geographical distribution of these records within Britain and Ireland (Fig. 7) is not dissimilar to the spring pattern (Fig. 4), with most on the English south coast. Relatively more were seen in south-east England, East Anglia and Scotland, however, and relatively fewer in the Midlands, Wales and south-west England (Table 1).

The first autumn arrivals in late July were in south-east England (Fig. 8) and in early August this was still the focal point, though occurrences were more widespread and also covered eastern England, East Anglia, the Midlands and south-west England. By mid-August the west, including Wales, began to figure more prominently. There were increasing numbers in south-west England by early September, but also further arrivals in East Anglia and south-east England and, completely separated from those in the south, in northern Scotland as well. Records declined after mid-September, but this was the time when autumn Hoopoes occurred in Ireland and the peak in north Scotland was at the end of September, much later than the peaks elsewhere.

Even though the English south coast accounted for over half of the records in autumn (as in spring), a greater proportion was on the east coast in autumn than in spring (Table 1). The pattern suggests a

Table 1. Comparison by regions of spring and autumn numbers of Hoopoes *Upupa epops* in Britain and Ireland during 1958–67

	Percentage distribution of spring records	Percentage distribution of autumn records	Proportion of spring to autumn
South-east England	22%	26%	2·7
South-west England	35%	28%	4·2
Midlands	11%	8%	4·3
Wales	9%	7%	4·5
East Anglia	9%	12%	2·4
Eastern England	3%	2%	3·7
North-west England	1%	2%	2·6
North-east England	1%	1%	2·5
North of Ireland	1%	1%	2·5
South of Ireland	6%	5%	4·9
Southern Scotland	1%	3%	1·9
Northern Scotland	1%	5%	0·8
Britain and Ireland	100%	100%	3·3
Hampshire to Northumberland	35%	41%	2·7
Dorset to Co. Kerry	51%	42%	4·2

simple westerly displacement of birds moving south in autumn, the numbers involved being far fewer than those in spring when Britain and Ireland lie directly in the path of birds overshooting on northward migration. The late autumn peak in north Scotland involved 14 individuals in the ten years and, although a similar concentration was found for 1839–1938 by Glegg (1942), this has not been satisfactorily explained.

Golden Oriole
Oriolus oriolus

A total of 257 Golden Orioles was recorded in Britain and Ireland during 1958–67, an average of 26 per year. The majority were between mid-April and mid-July, with 62% falling within the four weeks from 7th May to 3rd June (Fig. 9). At other seasons Golden Orioles were very rare vagrants at an average of little more than one per year. The sex was recorded for only 71%, of which 73% were males and 27% females. The adult male is, of course, very much more conspicuous than the female and, further, some females have black-and-yellow male-type plumage (England 1971), so these figures may not be entirely representative of the proportions of the sexes occurring. On the other hand, it is at least possible that some of the 'females' recorded were in fact first-year males. Even though the proportion of males to females may have been exaggerated, the records clearly show that the majority of females arrived in Britain and Ireland ahead of the majority of males (Fig. 9). This is the reverse of what might have been expected, for on spring migration in north Africa and Continental Europe the males precede the females by about ten days (Bannerman 1953). The sex of six out of the ten autumn (August–November) individuals was recorded, and all were males. The two March records both referred to males in 1966, in Huntingdonshire and Northamptonshire; in view of the unusually early dates, it seems quite probable that these were escaped cage-birds or even the same escaped individual.

More than two-thirds of the Golden Orioles seen in Britain and

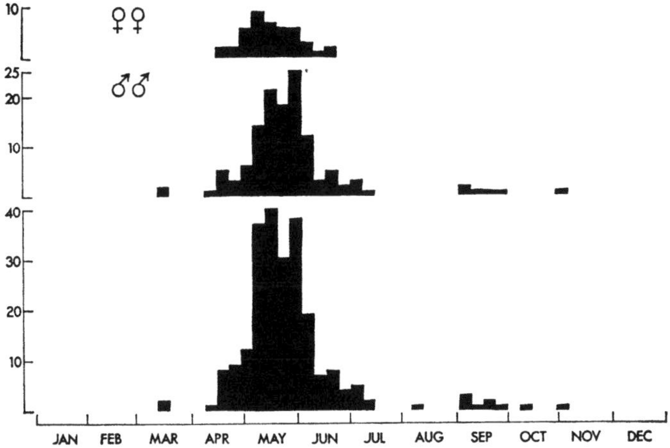

Fig. 9. Seasonal pattern of Golden Orioles *Oriolus oriolus* in Britain and Ireland during 1958–67 with the females, males and total records shown separately

Ireland during 1958–67 were in the last four years of the period (Fig. 10), with peaks in 1964 and 1967, and the pattern is quite different from that of the Hoopoes (compare Figs. 3 and 10). This is rather surprising, since the occurrences of both species are usually attributed to the same weather conditions (e.g. Davis 1964) and a 'good' year for one might be expected to be a 'good' year for the other. The periods of passage are rather different, of course—Hoopoes being mainly from the end of March to mid or late May and Golden Orioles from early May to early June—but there is still little correlation between the occurrences of the two species during the four weeks 7th May–3rd

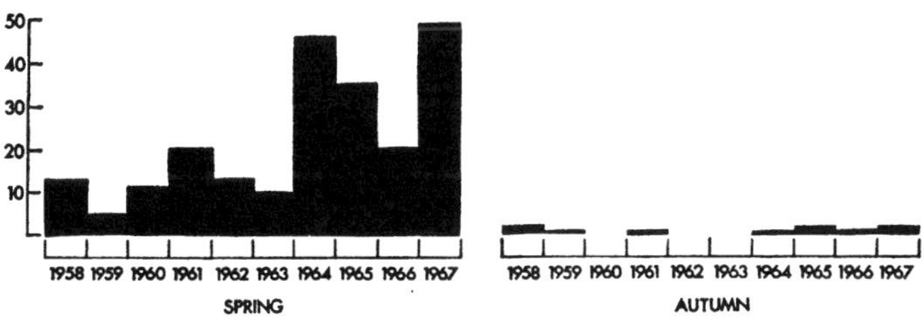

Fig. 10. Annual pattern of Golden Orioles *Oriolus oriolus* in Britain and Ireland during 1958–67 with the spring and autumn records shown separately

June which cover the main Golden Oriole passage and still significant numbers of Hoopoes (Table 2).

Table 2. Annual totals of Hoopoes *Upupa epops* **and Golden Orioles** *Oriolus oriolus* **in Britain and Ireland during 7th May–3rd June 1958–67**

Year	Hoopoe	Oriole	Year	Hoopoe	Oriole
1958	28	6	1963	22	2
1959	21	5	1964	22	29
1960	40	9	1965	28	27
1961	8	11	1966	19	14
1962	21	6	1967	19	36

The European breeding distribution of the Golden Oriole (Fig. 11) is very similar to that of the Hoopoe (Fig. 5), though the former has recently increased in Denmark and since 1944 extended its breeding range into southern Sweden, this being linked with the rise in mean spring temperatures in northern Europe (Voous 1960). The geographical distribution of the Golden Orioles within Britain and Ireland (Fig. 12) was also somewhat similar to that of the Hoopoes (Fig. 4), but by far the most were recorded on the Isles of Scilly. Significant numbers also occurred in Co. Cork, the English south coast counties from Devon to Kent, Essex, Suffolk, Huntingdonshire, Lancashire, and Orkney and Shetland. Although the majority were in coastal counties, it is noteworthy that this species, like the Hoopoe, is by no means solely a coastal vagrant and only two Midland counties (Derbyshire and

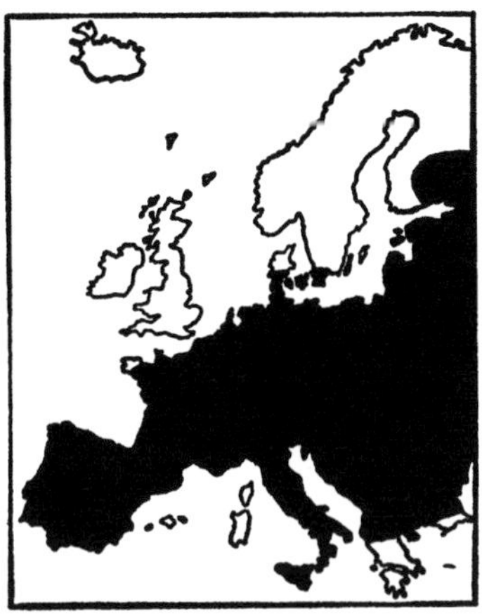

Fig. 11. European distribution of Golden Orioles *Oriolus oriolus* with the breeding range of this summer visitor shown in black (reproduced, by permission, from the 1966 edition of the *Field Guide*)

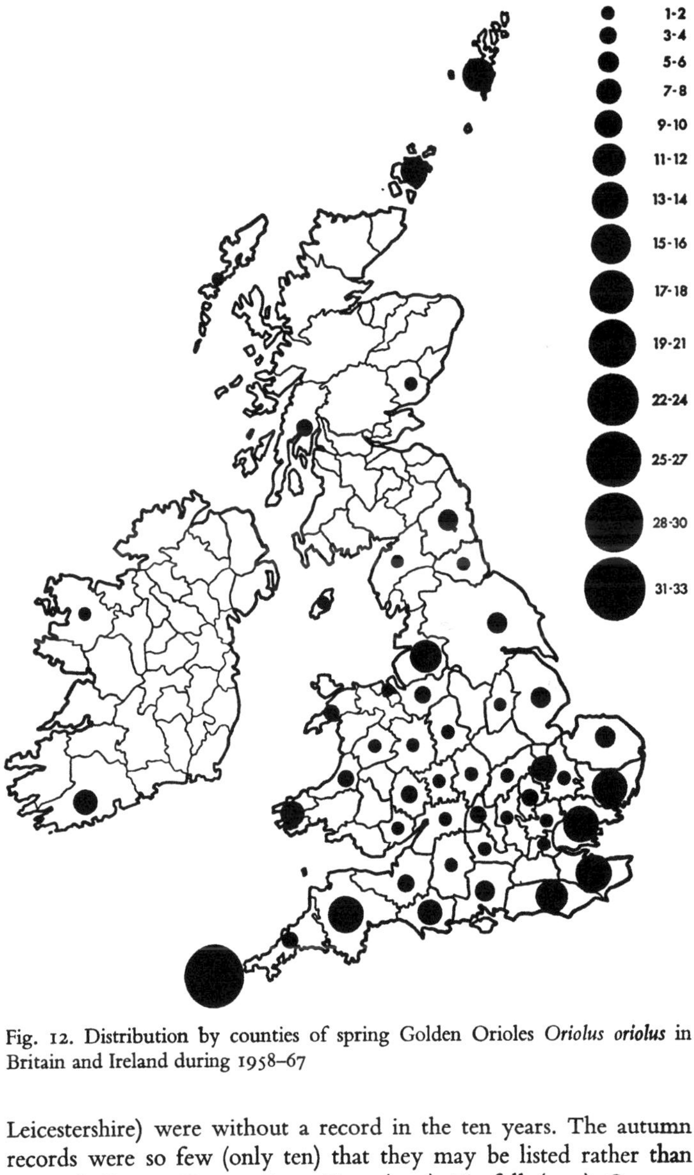

Fig. 12. Distribution by counties of spring Golden Orioles *Oriolus oriolus* in Britain and Ireland during 1958–67

Leicestershire) were without a record in the ten years. The autumn records were so few (only ten) that they may be listed rather than mapped: Devon, Dorset, Essex, Kent (two), Norfolk (two), Caernarvonshire, Argyll and Co. Cork.

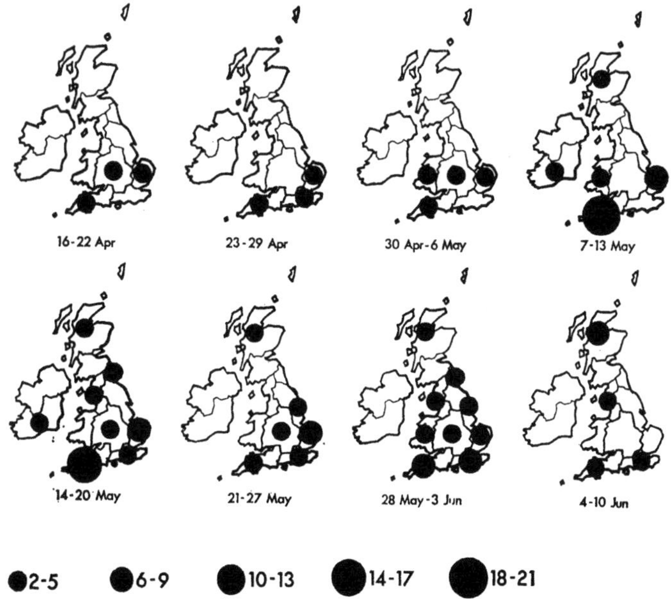

Fig. 13. Regional distribution by eight seven-day periods of spring Golden Orioles *Oriolus oriolus* in Britain and Ireland during 1958–67

The early spring arrivals from mid-April to early May were mainly in the south coast counties, East Anglia, the Midlands and, in the latter part of this period, Wales (Fig. 13). By the second week of May, the start of the four-week peak for this species, the largest numbers were in south-west England, but many were also recorded in East Anglia and some reached south Ireland and north Scotland at this time. The pattern was similar in mid-May, but by the end of May there was a more easterly bias, with a smaller proportion in south-west England. By 28th May–3rd June, the end of this four-week peak, Golden Orioles were more widely distributed than at any other time, with records in every region of England, Wales and northern Scotland. In early June they occurred most commonly in northern Scotland. This general pattern is similar to that of the Hoopoes.

It is interesting that the proportion of females to males in western areas (Dorset to Co. Cork) is double that in eastern England (Hampshire to Northumberland), the respective percentages being 36% and 18%. This is, of course, linked with two facts which have already been demonstrated: (1) the earlier arrival of females compared with males, and (2) the earlier peak in the south and west compared with the north and east.

Breeding records in Britain have been fewer this century than last, but there has been a slight increase in the last 20 years (Parslow 1973). In the ten years 1958–67 breeding was proved in two counties (Lancashire 1958 and probably 1959–61, Shropshire 1964) and probably occurred in another six (Essex 1958, Bedfordshire about 1959, Huntingdonshire 1961, Cardiganshire 1964, Sussex 1965 and 1966, Suffolk 1967). Males, and sometimes pairs, in suitable habitats in other counties in summer suggest that breeding may be less infrequent than appears from these twelve records.

It is surprising that, while both have similar breeding distributions on the Continent, 23% of the Hoopoes and yet less than 4% of the Golden Orioles occurred in autumn. This may be partly due to the differing wintering quarters of the two species. Hoopoes winter mainly north of the Equator in both east and west Africa, while Golden Orioles largely winter in tropical east Africa from Kenya and Uganda southwards (Vaurie 1959–65). The more easterly winter distribution of Golden Orioles presumably results in a more easterly standard direction on autumn migration and a consequent reduction in western vagrancy to Britain and Ireland.

Tawny Pipit
Anthus campestris

A total of 111 Tawny Pipits was recorded in Britain and Ireland during 1958–67, an average of eleven per year. Ages were reported for only eight adults and six immatures. The occurrences fell mainly into two periods—mid-April to mid-June and late August to early November—with most in September (Fig. 14). In contrast to the Hoopoes and Golden Orioles, there were more in autumn (88%) than in spring (12%). The European breeding distribution of the Tawny Pipit (Fig. 15) is very similar to that of the other two species (Figs. 5 and 11) and this radical difference in timing is therefore rather strange.

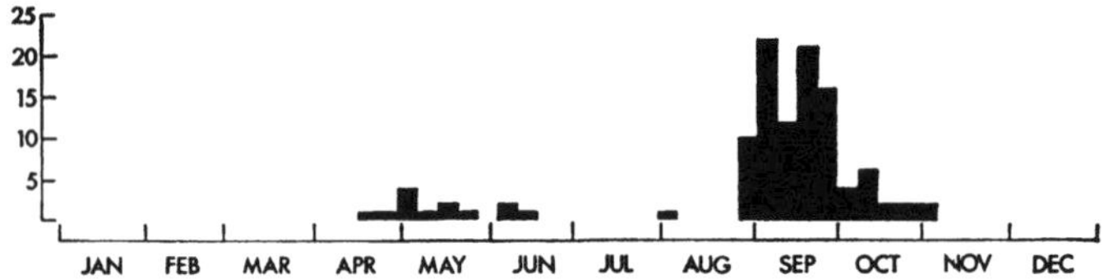

Fig. 14. Seasonal pattern of Tawny Pipits *Anthus campestris* on Britain and Ireland during 1958–67

The spring observations in the ten years were so few (13) that they may be listed: Cheshire, Cornwall, Isles of Scilly (two), Kent (three), Norfolk (two), Yorkshire (two), Fife and Shetland. Note that there was not a single spring record from Sussex which, back in 1905 and possibly again in 1906, produced the only British records of Tawny Pipits breeding, though these are no longer regarded as valid (Nicholson and Ferguson-Lees 1962). Turning to autumn, *The Handbook* was able to note over 40 records from Sussex, but otherwise only seven or eight in six or seven other counties from roughly 1868 to 1938. Although Sussex did not have quite such a monopoly in the ten years, over a quarter of the autumn records were in that county, and Dorset, Sussex, Kent and Suffolk together accounted for no less than 67% (Fig. 16). It is interesting that the only two autumn records north of 53°N were in Fife and the Outer Hebrides, the well-watched English east coast counties north of Norfolk not producing a single one. None

Fig. 15. European distribution of Tawny Pipits *Anthus campestris* with the breeding range of this summer visitor shown in black (reproduced, by permission, from the 1966 edition of the *Field Guide*)

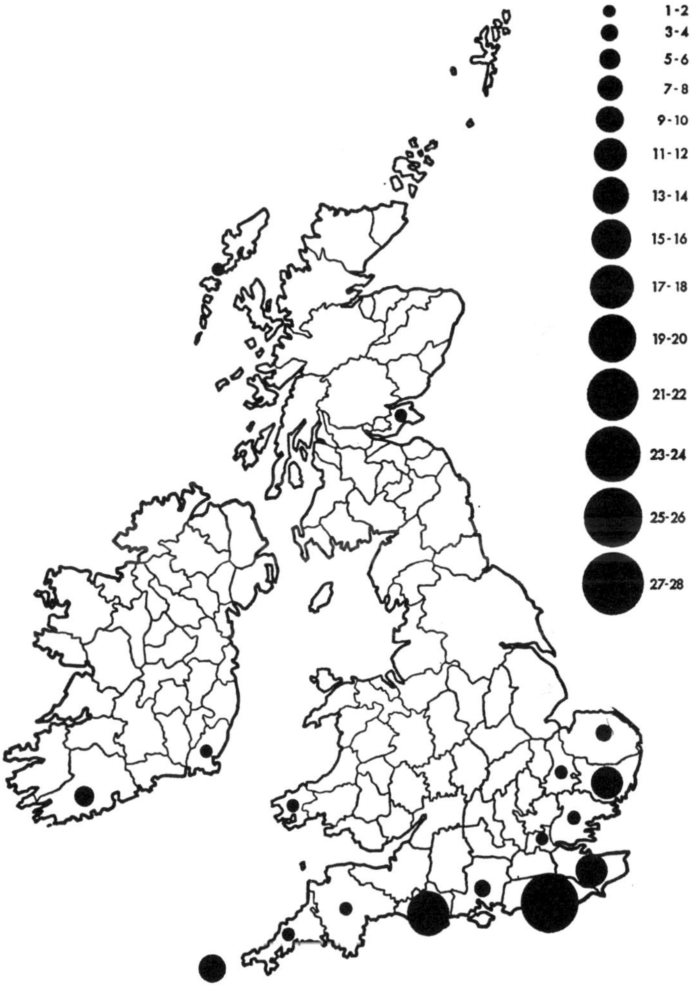

Fig. 16. Distribution by counties of autumn Tawny Pipits *Anthus campestris* in Britain and Ireland during 1958–67

of the 60 records in south-east England and East Anglia was after the first week in October, but seven of the 29 in south-west England, three of the six in the south of Ireland and both the Scottish records were in this late autumn period.

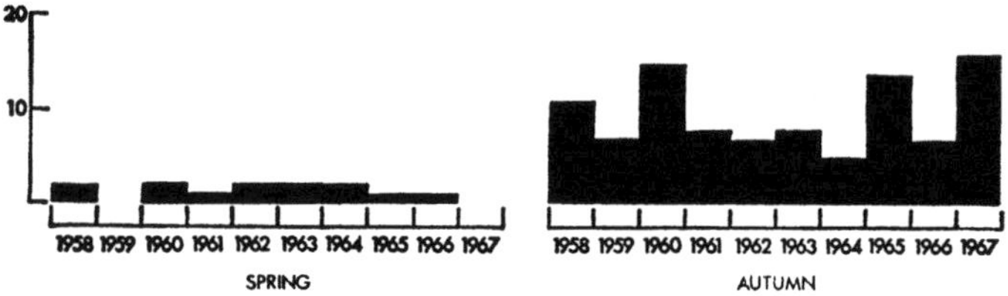

Fig. 17. Annual pattern of Tawny Pipits *Anthus campestris* in Britain and Ireland during 1958–67 with the spring and autumn records shown separately

There were just one or two spring records in eight of the ten years, while numbers fluctuated more widely in autumn (Fig. 17). The autumn peaks from the largest downwards were in 1967, 1960 and 1965, that in the last-named year being largely due to Suffolk records in 'the great immigration' of early September 1965 (Davis 1966). There has been a marked increase in frequency since the end of the ten years under review, with an average of 28 per year in 1968–72, compared with 11 per year in 1958–67.

The great scarcity of spring Tawny Pipits, compared with Hoopoes and Golden Orioles (which have similar European breeding distributions), suggests that this species is far less prone to overshooting on spring passage and, therefore, that it may be better equipped in some way to determine distance during migration than are the other two. Similarly, the concentration of autumn records in the south-east corner of England (Norfolk to Dorset) and absence of east coast records north of Norfolk suggests that the Tawny Pipit is less liable to lateral displacement by adverse weather, either through better navigational ability or because movement is not undertaken in such conditions: attributes one might expect to find in a largely diurnal migrant.

CHAPTER 2

Melodious Warbler, Icterine Warbler and Woodchat Shrike

This chapter considers two warblers, a species pair with almost mutually exclusive breeding ranges in Europe, and an unrelated species, a shrike, with a breeding range comparable to one of the warblers (Figs. 20, 27 and 31).

Melodious and Icterine Warblers are very similar in appearance and, indeed, have probably developed as separate species only relatively recently, on an evolutionary time-scale. Smaller and slimmer than Robins *Erithacus rubecula*, they are basically olive-green above and pale yellow below, with wide bills typical of the genus *Hippolais*. Separable in the hand by careful measurements, they were once considered almost inseparable when seen in the field, though Icterines usually have a pale wing-panel. It is now known, however, that they can be safely distinguished in the field by detailed observation of such features as the length of the wings at rest, the spacing of the folded primary feathers, the shape of the wings in flight and behaviour (Wallace 1964; Sharrock 1965a).

Woodchat Shrikes, unlike the two warblers in this chapter, are distinctive birds quite likely to be noticed as something unusual by the casual observer. About the same size as Yellowhammer *Emberiza citrinella* or Skylark *Alauda arvensis*, they have, however, a quite different 'jizz' (stance and general appearance). Like other shrikes, they perch rather upright on prominent vantage points and pounce upon their prey, such as beetles, bumble-bees or grasshoppers, which they may impale on thorns or the barbs of a barbed-wire fence, forming a 'larder'. Woodchats are strikingly patterned, with black wings and tail, white underparts, wing-patches and rump, and a chestnut-red crown. Juveniles are greyish-brown and barred, rather similar to the young of other shrikes, but they usually show pale traces of the adults' rump and scapular patches.

Melodious Warbler

Hippolais polyglotta

A total of 646 Melodious or Icterine Warblers was recorded in the ten years and almost 82% of these were specifically identified as one or the other. Melodious Warblers made up 41% of the identified birds, 217 being reported. The vast majority of these were in autumn (August to October) and only eight in spring (mid-May to June). The peak was during 27th August–16th September, with 53% of the autumn records falling within this three-week period (Fig. 18).

The majority of the few spring records were from Pembrokeshire, with others in Kent, Lancashire and Shetland. According to *The Handbook*, the breeding season in southern Europe is from the second half of May and in France from the last third of May. The small number (less than one per year) and lateness (more than half in June) of the spring records suggest that these were non-breeding individuals.

The autumn records were entirely in the south and west, with none on the English or Scottish east coasts north of Essex (Fig. 19). Six western counties (Co. Cork, Co. Wexford, Caernarvonshire, Pembrokeshire, Isles of Scilly and Dorset) accounted for no less than 74% of the autumn records in the ten years. Such a distribution can clearly be linked with the European breeding range of the species (Fig. 20) and this was discussed by Williamson (1959), who used the term 'drift-

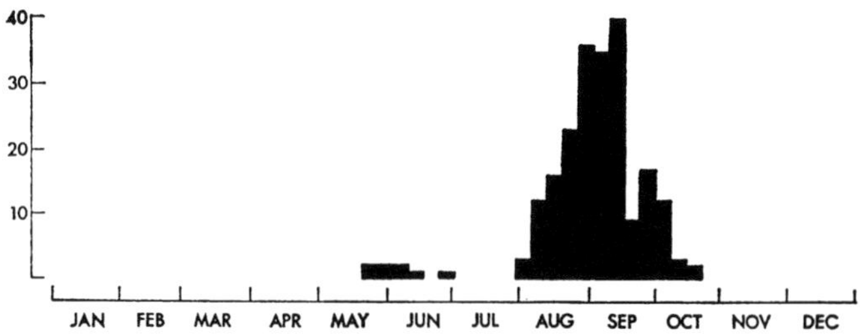

Fig. 18. Seasonal pattern of Melodious Warblers *Hippolais polyglotta* in Britain and Ireland during 1958–67

shadow' to describe it. The occurrence of relatively large numbers of this species, which has a southerly standard direction in autumn, on the south coasts of western England, Wales and Ireland (due north of the breeding range) could be explained by displacement in strong southerly

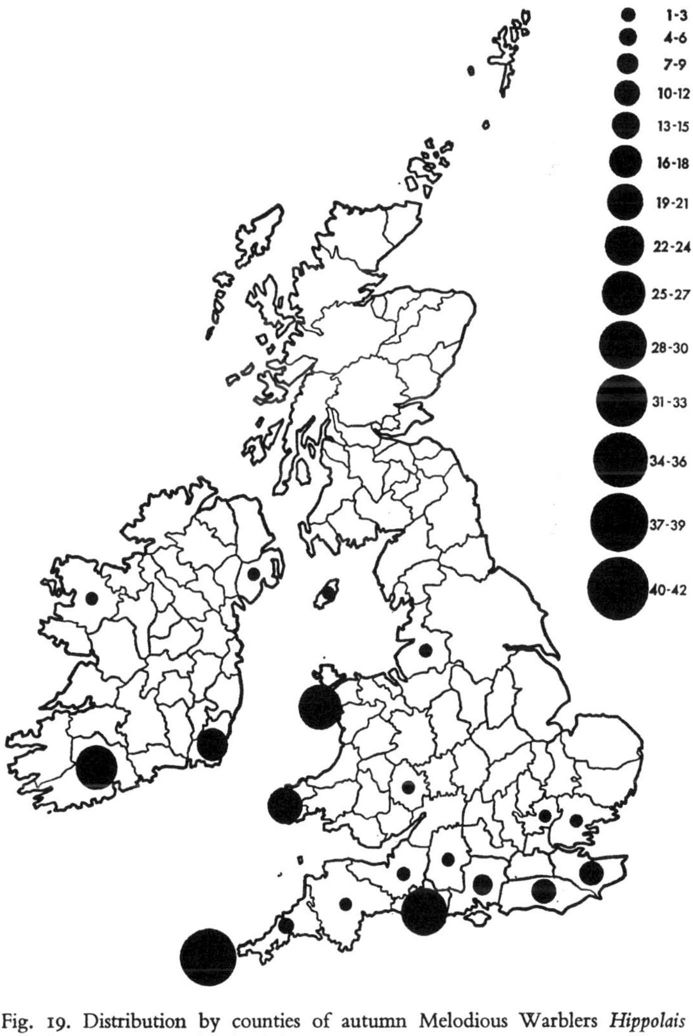

Fig. 19. Distribution by counties of autumn Melodious Warblers *Hippolais polyglotta* in Britain and Ireland during 1958–67

Fig. 20. European distribution of Melodious Warblers *Hippolais polyglotta* with the breeding range of this summer visitor shown in black (reproduced, by permission, from the 1966 edition of the *Field Guide*)

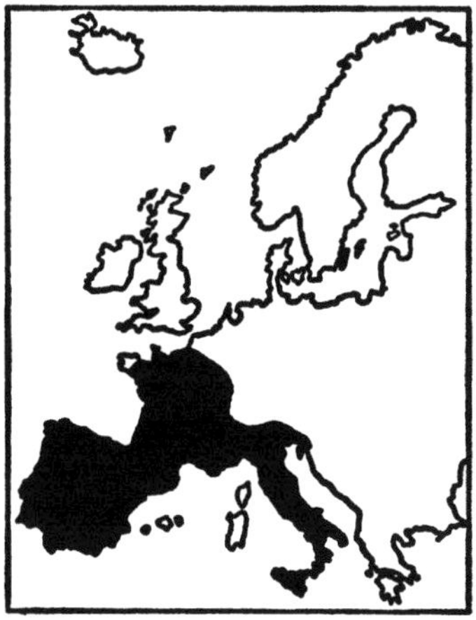

or south-easterly winds, by random post-breeding dispersal or by reverse migration. The experience of observatory workers is, however, that the species is not particularly associated with major falls of night-migrants in south-easterly frontal conditions, but occurs equally in anticylonic conditions with light winds. The mirroring of the breeding distribution in the pattern of vagrant records could suggest that dispersal is not random and, indeed, accords with the hypothesis that the British and Irish occurrences are largely the result of reverse migration, but this question is discussed further under Icterine Warbler.

The number of Melodious Warblers recorded each autumn has varied widely, from six in 1959 to 48 in 1962 (Fig. 21). At the time of *The Handbook*, only five records could be quoted and two of these are now considered to be dubious (Nicholson and Ferguson-Lees 1962). The increase in recent years has certainly been at least partly due to

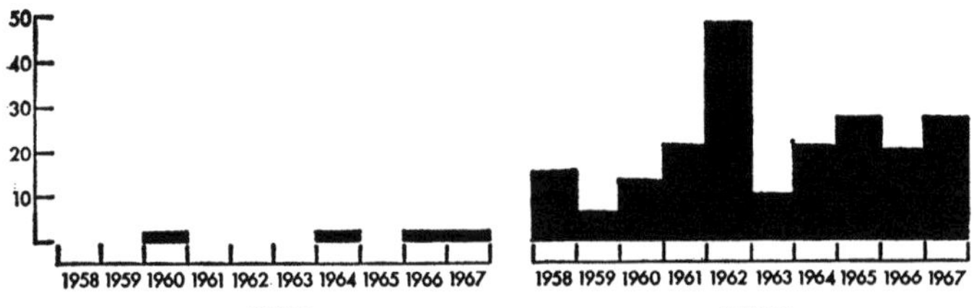

Fig. 21. Annual pattern of Melodious Warblers *Hippolais polyglotta* in Britain and Ireland during 1958–67 with the spring and autumn records shown separately

improved trapping and field-identification methods (Williamson 1959, Ferguson-Lees 1965). Williamson (1963) considered, however, that it might also be a function of population pressure, and particularly suggested that the exceptional number of occurrences in 1962 had the appearance of an irruption following a good breeding season. He argued that birds of similar range (Hoopoe, Aquatic Warbler, Woodchat Shrike and Tawny Pipit) had all been in fewer numbers than in other recent years, suggesting that an unusual meteorological situation was not the explanation, and he also pointed out that adults as well as young birds were involved. Now that all the records have been

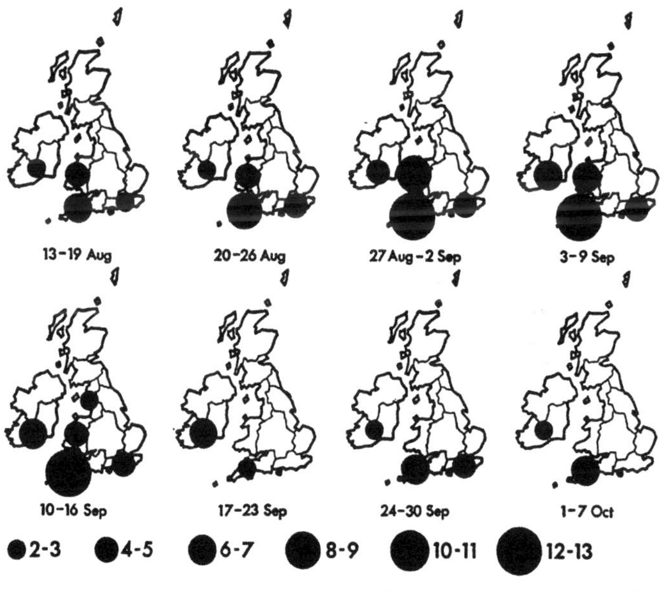

Fig. 22. Regional distribution in eight seven-day periods of autumn Melodious Warblers *Hippolais polyglotta* in Britain and Ireland during 1958–67

collated, the comparison with the other four species named by Williamson still holds good (23% of the 1958–67 autumn records of Melodious Warbler were in 1962, compared with only 7% for Tawny Pipit, 5% for both Hoopoe and Aquatic Warbler, and 3% for Woodchat Shrike). It is not possible, however, to confirm the occurrence of adults with first-year birds for, regrettably, county reports seldom record the age data which may be obtained by observatory workers. Indeed, the age is given for only 13% of the Melodious Warblers in the ten years, 89% of them being first-year.

In the ten years the autumn arrival of this species was almost synchronous throughout the limited area of its vagrancy in Britain and Ireland (Fig. 22), although it did seem to be slightly earlier in the east than in the west. The peak in south-east England was 20th August–16th September, in Wales and south-west England 27th August–16th September, and in the south of Ireland 3rd–23rd September.

This species is also discussed in the following section.

Icterine Warbler
Hippolais icterina

A total of 311 Icterine Warblers was recorded in the ten years, approximately 59% of the Melodious or Icterine Warblers which were specifically identified. As with the Melodious Warblers, by far the majority (85%) were in autumn, but spring records considerably exceeded those of the more westerly species (47, compared with only eight Melodious). Almost half of the spring records occurred in one aberrant year (1967), but there were records annually in autumn (mainly August and September) with a very marked peak (51% of the autumn records) in the fortnight 27th August–9th September (Fig. 23).

The spring records were mainly in Shetland (31), with scattered records of one or two in the ten years in Co. Cork, Co. Wexford, Pembrokeshire, Dorset, Sussex, Yorkshire, Fife and Angus, and three in both Caernarvonshire and Norfolk. Most of the Shetland spring records came in 1967 when, from 25th May to 5th June, about 20 Icterine Warblers occurred on Fair Isle (Dennis 1967) and there were three elsewhere at about the same time (Co. Cork, Sussex and Angus). Even disregarding this exceptional influx, however, Shetland still accounted for almost half of the British and Irish spring records. In this connection, note that eleven of the records given for Scotland in *The Handbook* were in spring, compared with only nine in autumn.

The autumn pattern (Fig. 24) was very different from that in spring, with most records on the British east coast, especially (in decreasing order) in Norfolk, Northumberland, Lincolnshire, Yorkshire, Shetland and Fife. There were also significant numbers in three south coast counties (Kent, Dorset and Isles of Scilly) and a very striking concentration in Co. Cork, which produced as many as the much more fully covered county of Norfolk (*cf.* Fig. 1). This peculiar situation will be discussed later.

The annual totals in spring and autumn are shown in Fig. 25. Spring numbers varied from one to three each year, apart from the exceptional number in 1967 already noted and six in 1964. Of these latter, five were in the seven days 28th May–3rd June, in Caernarvonshire, Norfolk, Yorkshire and Shetland (Ferguson-Lees 1964). The autumn totals varied less, with the lowest in 1961 (16) and the highest in 1966 (52). These numbers were considerably more than had been recorded previously, probably due at least in part to the increase in the number of observers, but it is noteworthy that the highest annual total up to 1953 was only eleven in 1951 (Ferguson-Lees 1954), compared with the average of 31 over the ten years.

The region of peak passage on the British east coast became progressively more southerly through the autumn (Fig. 26). The first peak was in northern Scotland (20th–26th August), then southern Scotland, north-east England and eastern England (27th August–2nd September), followed by East Anglia (3rd–9th September) and then lastly the small number in south-east England (3rd–23rd September). It should be noted that, whereas almost half of the East Anglian Icterine Warblers were concentrated in a single seven-day period, the exactly equal

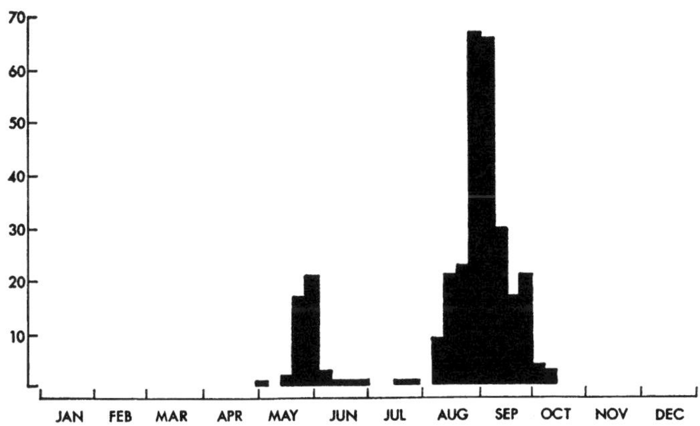

Fig. 23. Seasonal pattern of Icterine Warblers *Hippolais icterina* in Britain and Ireland during 1958–67

number recorded in the south of Ireland was spread out relatively evenly over the whole eight weeks (though reaching a peak during 27th August–2nd September).

The European breeding distribution of the Icterine Warbler (Fig. 27) helps to explain the pattern of records in Britain and Ireland. The species has a standard direction west of north in spring and east of

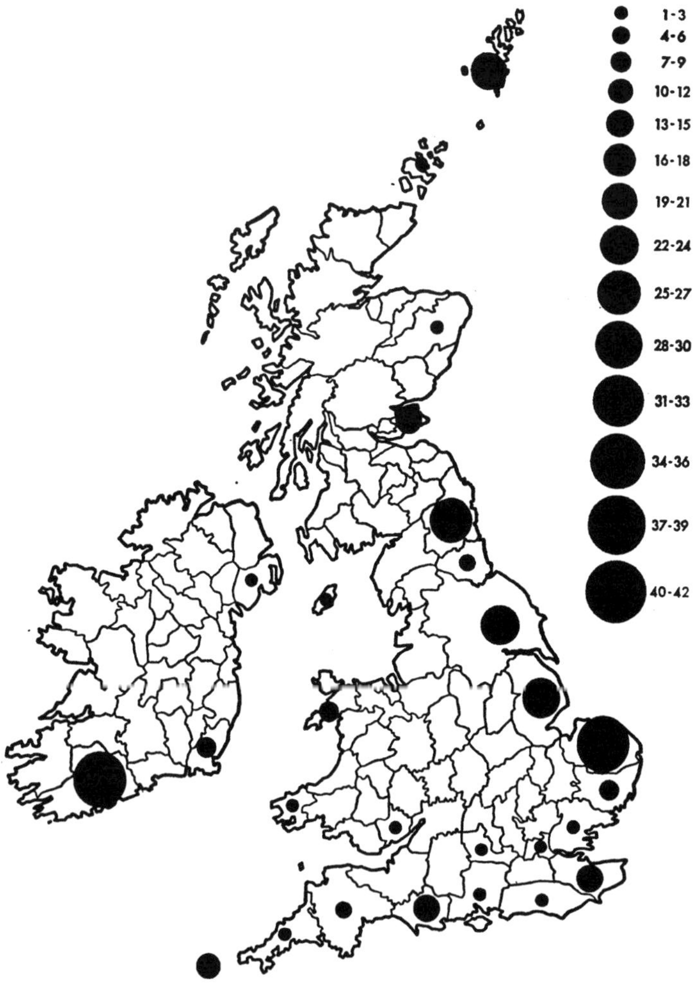

Fig. 24. Distribution by counties of autumn Icterine Warblers *Hippolais icterina* in Britain and Ireland during 1958–67

south in autumn. The concentration of spring records in Shetland (66%, or 46% if the 1967 ones are excluded) suggests that it is largely the Scandinavian or north-east European parts of the population which tend to overshoot (or which are diverted westwards in exceptional weather conditions) in spring. The progressively more southerly peaks down the British east coast in autumn are somewhat reminiscent of the

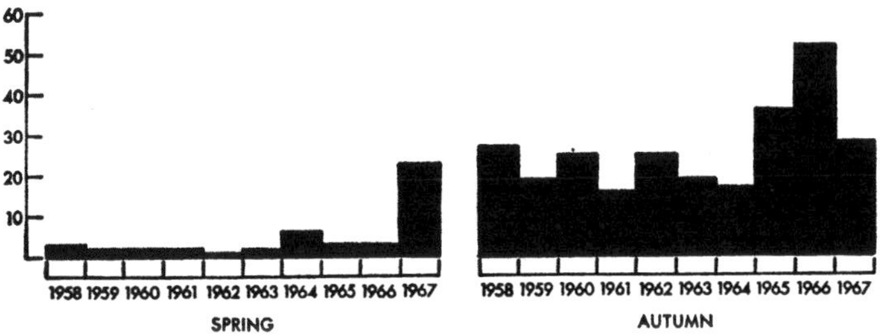

Fig. 25. Annual pattern of Icterine Warblers *Hippolais icterina* in Britain and Ireland during 1958–67 with the spring and autumn records shown separately

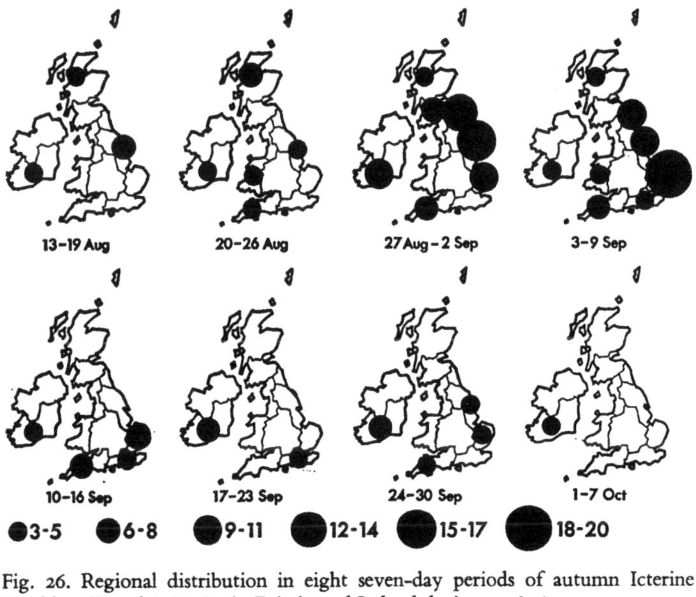

Fig. 26. Regional distribution in eight seven-day periods of autumn Icterine Warblers *Hippolais icterina* in Britain and Ireland during 1958–67

situation in the Barred Warbler discussed by Davis (1967) when postulating that vagrancy in that species was directed north-westwards and not at random. The situation is not identical, however, for (unlike Barred Warblers) Icterine Warblers are not concentrated in Shetland in autumn and numbers do not decrease as one progresses southwards down the east coast. Nevertheless, the phased timing is similar and might be taken as suggesting a common cause. As autumn migration takes place, the bulk of the European population will be at progressively more southerly points and this is likely to be reflected in the pattern of vagrancy to the west (a similar situation has been demonstrated in the spring Hoopoe records—Fig. 6—with progressively later peaks northwards up the east coast). During the autumn, the tracks of Atlantic depressions across Britain become progressively more southerly and, since vagrancy of Icterine Warblers on the east coast is closely associated with large arrivals of grounded night migrants of Continental origin, this may also be a contributory cause of the phenomenon. Lack (1960) has shown, however, that many big or moderate arrivals on the east coast occur with neither fog nor total cloud and, therefore, it seems more likely that the simpler explanation (that vagrancy merely mirrors the latitude of the bulk of the population at any particular time) is the one which is largely correct.

Turning now to the west of Britain and Ireland, there is the puzzling situation, already mentioned, that the total of Icterine Warblers in Co. Cork was equal to that in Norfolk (and in the south of Ireland was equal to that in East Anglia). Williamson (1959) drew attention to the

Fig. 27. European distribution of Icterine Warblers *Hippolais icterina* with the breeding range of this summer visitor shown in black (reproduced, by permission, from the 1966 edition of the *Field Guide*)

drift-shadow pattern in the records of Icterine and Melodious Warblers, reflecting their European breeding distributions, with, for instance, all but two of 'the score or so Icterine Warblers [in 1958] confined to the east coast between Fair Isle and Norfolk, whilst a similar number of Melodious Warblers were confined to the south coast and the Irish Sea basin'. Later, however, the same author (1960, 1963) noted that the drift-shadow situation was departed from in 1960 and subsequently, with as many Icterine as Melodious Warblers at the Irish Sea observatories and, in some years, more Icterine Warblers in the west than on the east coast. He suggested that this change in pattern 'may be another case where we are witnessing the reaction of a successful part of the population, the central and east European, to high atmospheric pressure after the breeding season'. From a consideration of all the records, however, it seems to be far from proven that any change took place in 1960 and subsequently. In 1958–59, 18% of the Icterine Warblers were in south-west England, Wales and the south of Ireland, compared with 26% in these areas in 1960–67. But one station—Cape Clear Island, Co. Cork—was not manned until 1959 and then was responsible for 40% of the Icterine Warbler records in the west: if the records from there are eliminated from the calculations, the pre-1960 and 1960–67 percentages become 13% and 17% respectively, a change which is not significant. Thus, there is virtually no evidence of a change in vagrancy pattern.

Since some 18% of the Icterine and Melodious Warblers seen in Britain and Ireland were not specifically identified, the patterns of the two species are best demonstrated by comparison. The proportions of both in the various regions are shown in Table 3. On the British east coast, 99% of those specifically identified were Icterine, as were 56% in Kent and 58% in Co. Cork. In all the other counties in the south of Ireland, Wales and southern England, Melodious outnumbered Icterine. The indeterminate records make some of these percentages dubious, but it is possible (by first assuming that the indeterminate birds were all Melodious and then that they were all Icterine) to calculate the maximum range of error. By this means, it can be shown that Icterine Warblers vastly outnumbered Melodious Warblers on the British east coast, were slightly in excess in Kent and were almost certainly in excess in Co. Cork; whereas Melodious Warblers were the more frequent in Co. Wexford, Caernarvonshire, Pembrokeshire, Dorset, Hampshire and Sussex. The situation in the Isles of Scilly, Cornwall and Devon is uncertain, though those specifically identified were mainly Melodious. It should be mentioned that, in areas where the Melodious Warbler is

Table 3. Numbers and proportions of autumn Icterine Warblers *Hippolais icterina* and Melodious Warblers *H. polyglotta* in various areas of Britain and Ireland during 1958–67

	Number of Icterine	Number of Melodious	Number not identified	% Icterine of identified	Possible extreme percentages (see text)
Co. Cork	35	25	11	58%	49–65%
Co. Wexford	9	17	1	35%	33–37%
Caernarvonshire	9	26	5	26%	23–35%
Pembrokeshire	3	21	10	13%	8–38%
Isles of Scilly	11	39	36	22%	13–55%
Cornwall and Devon	5	8	16	38%	17–72%
Dorset	15	28	6	35%	31–43%
Hampshire	1	8	6	11%	7–47%
Sussex	1	11	0	8%	8%
Kent	14	11	1	56%	54–58%
Essex to Shetland	140	1	19	99%	88–99%

known to occur, the indeterminate records are probably biased towards this species (which is marginally less easy to identify positively in the field); whereas on the British east coast, where it is virtually unknown, most indeterminate records are likely to be briefly-glimpsed Icterine Warblers (though there is the possibility that the occasional Melodious may be dismissed by careless observers as 'only another Icterine').

The large number and high proportion of Icterine Warblers in Co. Cork is exceedingly difficult to explain. Icterine and Melodious Warblers have very similar occurrence-patterns in autumn in the western areas (Fig. 28) and, indeed, often appear together. Throughout these western areas, both species have prolonged passage periods, unlike the very marked peaks of Icterine Warblers at east coast stations. The pattern of occurrence of the Melodious Warbler accords with what would be expected if the cause were reverse migration in a species with a southerly standard direction (see earlier section), but the

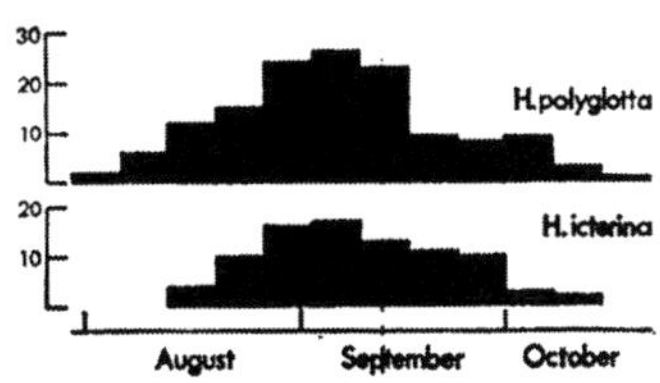

Fig. 28. Pattern in south Ireland, Wales and south-west England of August–October Melodious Warblers *Hippolais polyglotta* and Icterine Warblers *H. icterina* during 1958–67

pattern of western records of the Icterine Warbler, despite this species' more easterly distribution and standard direction east of south, is remarkably similar. It seems inescapable that the immediate origin of both species is the same and that the Icterine Warblers occurring in the western areas are ones which have moved westwards into northern France before wandering northwards or north-westwards into south-western Britain and the south of Ireland. It may well be, therefore, that the occurrence of both species in these western areas is primarily due to random post-breeding dispersal. Although only 8% of the autumn Icterine Warblers were aged and the information published (one adult and 19 first-year), most were probably first-year and 89% of the aged Melodious Warblers were also first-year. It thus seems likely that the primary influence on both species' occurrences in the west is random post-juvenile dispersal in anticyclonic conditions, but reverse migration may subsequently contribute to the patterns; the same is probably true also of many of the other scarce migrants and vagrants at the western observatories. This does not, however, explain why Co. Cork has received so many more Icterine Warblers than have other western counties and the solution to this discrepancy has yet to be suggested.

Woodchat Shrike
Lanius senator

A total of 125 Woodchat Shrikes was recorded in Britain and Ireland in the ten years. Unlike the five previous species (Hoopoe, Golden Oriole, Tawny Pipit, Melodious Warbler and Icterine Warbler), there were almost equal numbers at the two seasons, 66 in spring and 57 in autumn. At the time of *The Handbook*, there was a greater discrepancy, with more in spring. Improved field-identification of young birds in autumn has probably resulted in the truer picture evident today.

Spring occurrences were from mid-April to mid-June (mainly May and early June) and autumn ones from late July to early October (mainly late August to early September), with two records in early

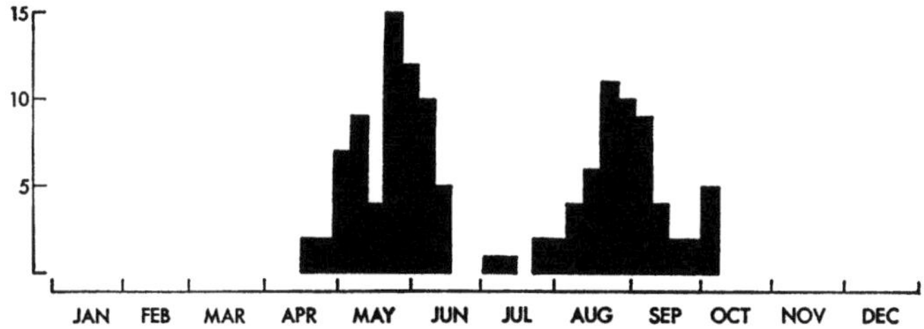

Fig. 29. Seasonal pattern of Woodchat Shrikes *Lanius senator* in Britain and Ireland during 1958–67

July (Fig. 29). Almost half of those in spring were sexed, resulting in a total of 26 males and only six females. In autumn, of 54 individuals which were aged, 42 were first-year and twelve adults (only five of the latter were sexed as three males and two females).

In spring most Woodchat Shrikes were seen in Norfolk, with smaller numbers in Lincolnshire, Pembrokeshire, Co. Wexford, Caernarvonshire, the Isles of Scilly and Shetland, at least one in each English south coast county and several additional east coast counties, and scattered records elsewhere (Fig. 30). The Woodchat Shrike's European breeding distribution (Fig. 31) is closely similar to that of the Melodious Warbler (Fig. 20), its strongholds being in Iberia, France, Italy, the Mediterranean islands and the western Balkans. Though it additionally extends further east, the populations east from the Low Countries, Germany and Switzerland are rather sparse (Ferguson-Lees 1965). It is, therefore, of interest that 53% of the Woodchat Shrikes occurred in spring, compared with less than 4% of the Melodious Warblers. The spring occurrences are usually attributed (along with records of other southern vagrants) to the overshooting of migrants in fine anticyclonic weather. If this were the simple explanation, however, one would expect most of the spring records to be concentrated in the English south coast counties (leading, perhaps, to a distribution not unlike that of the Hoopoe shown in Fig. 4). The relative scarcity of south coast records (only 21% compared with 57% for the Hoopoe) is thus somewhat surprising, as is the concentration in Norfolk, a county north of the species' breeding range and, indeed, with a mainly north-facing coastline (ten of the twelve records were from north-facing parts of the coast). The concentration of spring records in Norfolk may not be regular, however, for *The Handbook* merely noted 'over forty at intervals, mostly in S. and E., as far north as Norfolk' and one in 1955 was the only record in the Cley area before 1958 (Richardson 1962), compared with five in the subsequent ten years.

Autumn Woodchat Shrikes showed a greater bias towards the west and the majority of the records were in four counties—Isles of Scilly, Dorset, Pembrokeshire and Co. Wexford (Fig. 32). This is a very

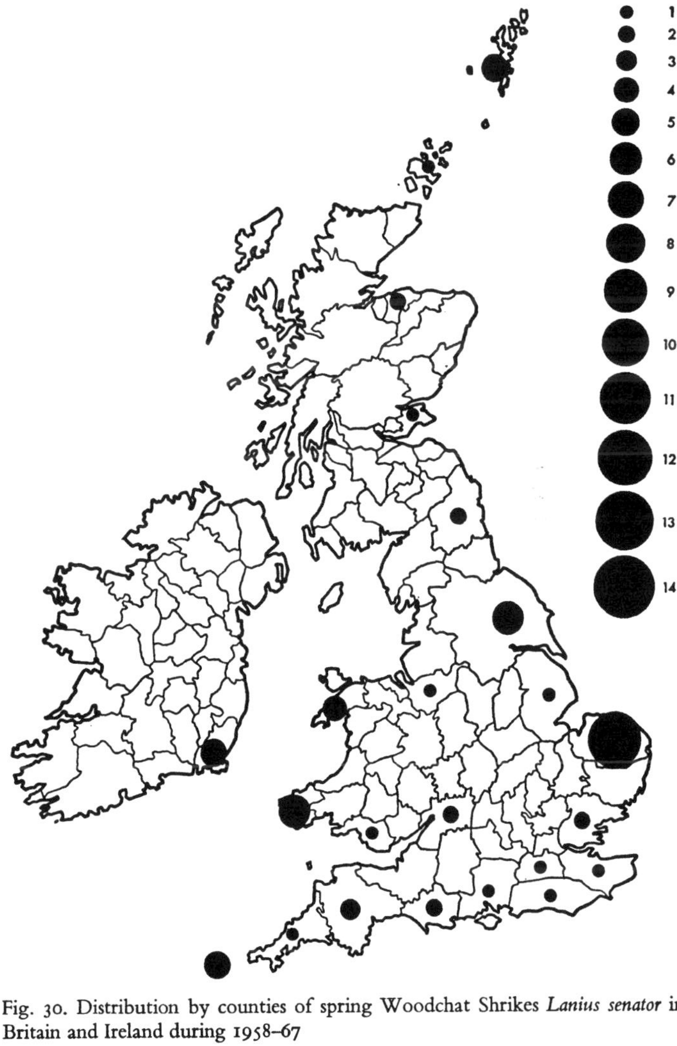

Fig. 30. Distribution by counties of spring Woodchat Shrikes *Lanius senator* in Britain and Ireland during 1958–67

similar distribution to that of the Melodious Warbler (Fig. 19), though (unlike that species) there were a few east coast records—no doubt a reflection of the slightly more easterly extended range of the Woodchat Shrike—and it is, perhaps, surprising that so few were recorded in Co. Cork.

The number of Woodchat Shrikes occurring in Britain and Ireland varied widely over the ten years, from one to 14 in spring and from one to 13 in autumn (Fig. 33). The coincident drop in both spring and autumn in the years 1961–63 called for considerable comment (e.g. Harber *et al.* 1964, Ferguson-Lees 1965). The fall in numbers was particularly noticeable because it came immediately after the exceptional spring of 1960, but nevertheless there was an average of only just over three per year during 1961–63, compared with slightly less than 18 per year in the three previous years, over 15 per year in the four subsequent ones and (outside the period under review) 13 per year in 1968–72. Other southern vagrants did not show a corresponding decline, so it seems possible that this was a reflection of a low level in the Woodchat population, though there is no direct evidence of this.

The breeding season starts in late April in Iberia and towards the end of May in central Europe, so that the spring peak in Britain (the last fortnight in May) is rather late. This, together with the excess of males recorded (26 to only six females), suggests that non-breeding birds are concerned, perhaps ones which have already overshot or wandered north of their European breeding range before crossing the North Sea to the British east coast, and that it is the eastern part of the population which is largely involved at this time. This is confirmed

Fig. 31. European distribution of Woodchat Shrikes *Lanius senator* with the breeding range of this summer visitor shown in black (reproduced, by permission, from the 1966 edition of the *Field Guide*)

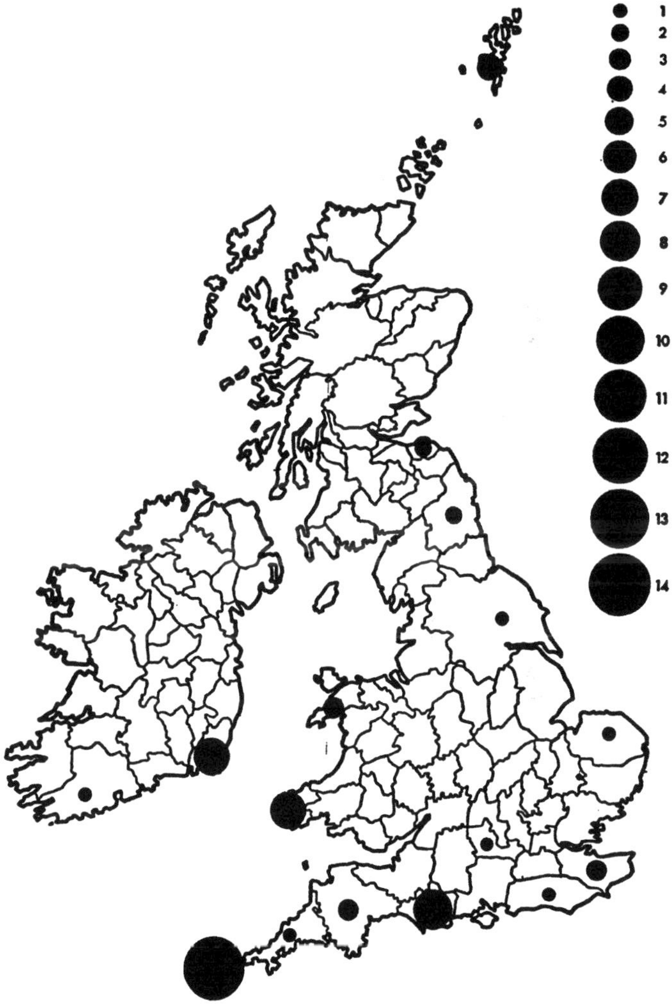

Fig. 32. Distribution by counties of autumn Woodchat Shrikes *Lanius senator* in Britain and Ireland during 1958–67

by the timing of passage in the various regions (Fig. 34). The distribution of spring records was double-peaked (Fig. 29), with a small peak in early May, a lull in mid-May and then the major peak at the end of May. The earliest ones, in April, were in Wales and those in early May

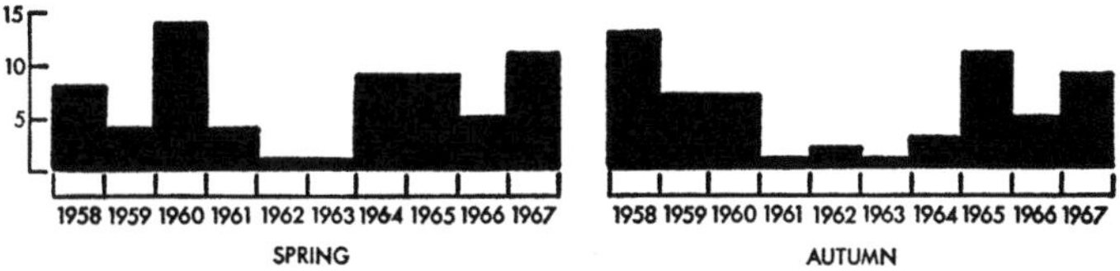

Fig. 33. Annual pattern of Woodchat Shrikes *Lanius senator* in Britain and Ireland during 1958–67 with the spring and autumn records shown separately

mainly in south-west England and south Ireland. The mid-May lull (14th–20th May, not shown in Fig. 34) produced only three in East Anglia and one in eastern England. The main peak in the latter part of May was widespread, though most records were in East Anglia, northern Scotland and south-west England. The peak in eastern England was in the last fortnight of the spring passage, when there were further (but fewer) East Anglian records and, surprisingly, a peak in Wales. The numbers involved were so small that it would be dangerous to base too much speculation upon them, but it is surely significant that during the first peak (16th April–13th May) 65% of the records were in the west and only 20% in the east, whereas during the second peak (21st May–17th June) 33% were in the west and 52% in the east. This pattern, with an earlier spring peak in the west than in the east, is similar to that already demonstrated in Chapter 1 for the Hoopoe and the Golden Oriole. The earlier records are probably the result of overshooting by Iberian birds, whereas those later (mainly in the east) are likely to be of more easterly and northerly origin.

The autumn records were so concentrated in the west (Fig. 32) that

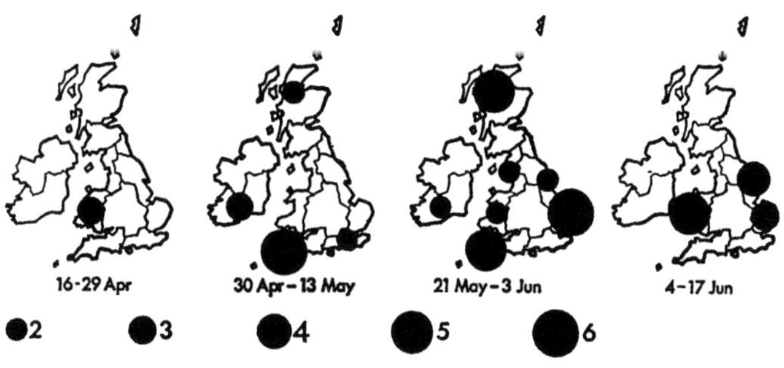

Fig. 34. Regional distribution in four 14-day periods of spring Woodchat Shrikes *Lanius senator* in Britain and Ireland during 1958–67. (Note that a seven-day gap during 14th–20th May is not depicted)

analysis of periods of passage in the various regions is unrewarding. It may be mentioned, however, that four of the eight east coast records came during 10th September–7th October (after the autumn peak), whereas only eight of the 43 west coast records were at this time. Nevertheless, the numbers are really too small to be significant. The similarity between the breeding ranges and patterns of autumn vagrancy of the Melodious Warbler and the Woodchat Shrike suggests that the occurrences of these species at the westerly stations is due to a common cause. This is all part of a larger problem, however, for the whole question of vagrancy at the Irish Sea stations in autumn is clouded by theory and counter-theory and deserves greater study. Radar evidence, which has helped to elucidate some English east coast problems, is lacking and, in any case, it is not large-scale movements which are puzzling: merely the disproportionate numbers of rare birds.

CHAPTER 3

Rough-legged Buzzard, Temminck's Stint and Long-tailed Skua

We now turn to three species with Arctic breeding distributions, the European portions of which are virtually confined to Fenno-Scandia and north Russia (Figs. 38, 41 and 51).

Rough-legged Buzzards are about the same size as the common Buzzards *Buteo buteo* of western and northern Britain. They are rather variable but usually have a dark belly, whitish underwing with dark carpal patches and a white tail with a broad black terminal band, the latter being the best single feature for identification (Christensen *et al.* 1971–73). To be found in Britain most often in areas of open country, such as reclaimed coastal pasture or heathland, they often hover when hunting.

Temminck's Stints are tiny waders the size of Robins *Erithacus rubecula*. They are similar to the commoner Little Stint *Calidris minuta*, but have white sides to the tail, pale legs and a different call and behaviour. Other confusingly similar small waders (stints and peeps) create identification problems for the non-expert (see Wallace and Grant 1974). Temminck's Stints occur most often along the wet margins of lakes and marshes

Long-tailed Skuas are less piratical in their habits than other skuas—seabirds that force gulls and terns to drop their food. Long-tailed Skuas are delicate birds, slender, with narrow, pointed wings, and are buoyant and agile in flight. They are about the same size as Black-headed Gulls *Larus ridibundus*, but the adults have elegant elongated central tail feathers, six to ten inches long; they are white below and grey-brown above, with a sharply demarcated black cap. Arctic Skuas *Stercorarius parasiticus* are similar but are more bulky, with a less slender and elongated appearance, and have shorter (3″–5″) tail projections. The immatures are more similar, but young Long-tailed are often greyer than young Arctics (see Bell 1965). Though occasionally occurring inland, they are most likely to be seen flying over the sea, especially in stormy weather.

Rough-legged Buzzard
Buteo lagopus

Prompted by exceptional numbers during the winter of 1966/67, Scott (1968) discussed in some detail the records of Rough-legged Buzzards in Britain, and this section can therefore be fairly brief. Rough-legged Buzzards are largely winter visitors to Britain and even those on passage are probably seen at several different localities as they move through the country. For these reasons, it is more difficult to determine the number of separate birds involved than it was in the cases of the other six scarce migrants already dealt with. It is probable, however, that rather less than 269 individuals were recorded in the ten years.

Records covered the period from mid-August to May, but by far the majority of first sightings were in October and November (Fig 35). Only three of the 42 seven-day periods between 13th August and 3rd June were without a record of a 'new' individual and, since it seems most unlikely that there were continuing arrivals throughout this time, it is clear that wintering birds wandered a good deal during their stay. This makes it exceedingly difficult to decide how many were involved each year, but the error from this source is likely to be fairly constant from year to year so that it is at least possible to obtain comparative figures. Since the period covered in this book is 1st January 1958 to 31st December 1967, full totals cannot be given for the 1957/58 and 1967/68 winters and so Fig. 36b shows the totals for each of the nine winters 1958/59 to 1966/67. This clearly demonstrates the exceptional nature of the influx in 1966/67, with 40% of all the records in that one winter. Williamson (1961) noted that there were more Rough-legged Buzzards in evidence during October and November 1960 than there had been for several years and that winter (1960/61) shows as the second highest peak in the nine years, but with fewer than half the number of 1966/67. The influx at this time was documented by Ferguson-Lees and Williamson (1960, 1961), whilst influxes in previous years were listed by Scott (1968) and *The Handbook*.

The number actually wintering in Britain each year is a matter for speculation. Scott (1968) defined wintering birds as individuals which remained in one area throughout December to February, 'or at least for periods of over four weeks'. On this basis, he estimated that at least 57 and probably 67 individuals wintered in Britain in 1966/67. Defining wintering individuals rather differently, as those staying in an area for at least two weeks, including periods in both of the two years (i.e. minimum requirements would be 19th December–1st January or 31st December–13th January), one arrives at a total of about 32 individuals wintering in Britain in 1966/67 and only eleven in the previous nine

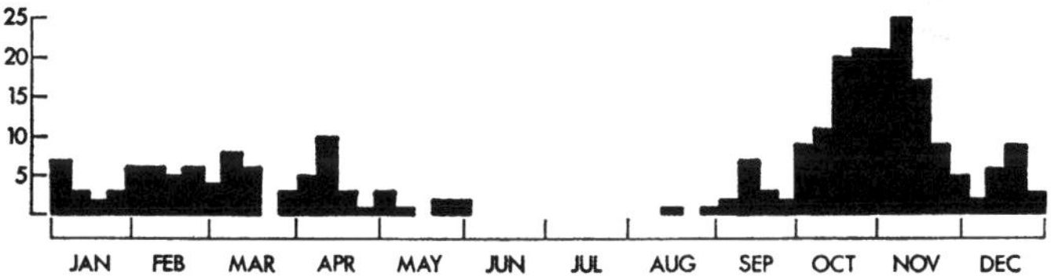

Fig. 35. Seasonal pattern of Rough-legged Buzzards *Buteo lagopus* in Britain and Ireland during 1958–67

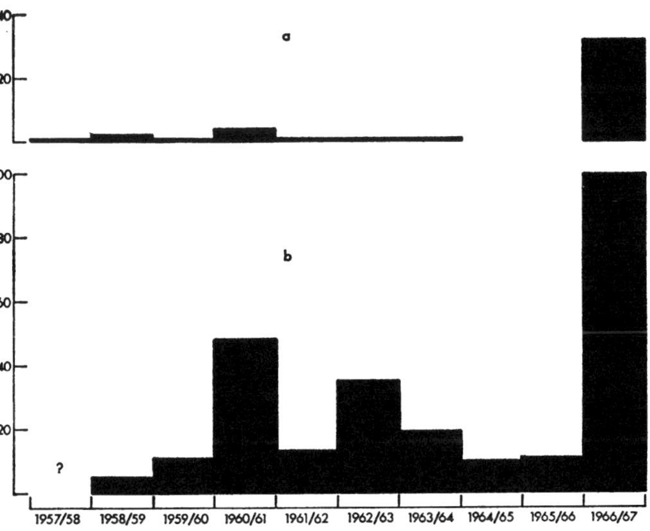

Fig. 36. Annual patterns in Britain and Ireland of (a) wintering Rough-legged Buzzards *Buteo lagopus* from 1957/58 to 1966/67 and (b) all Rough-legged Buzzards from 1958/59 to 1966/67

Fig. 37. Distribution by counties of wintering Rough-legged Buzzards *Buteo lagopus* in Britain and Ireland from 1957/58 to 1966/67

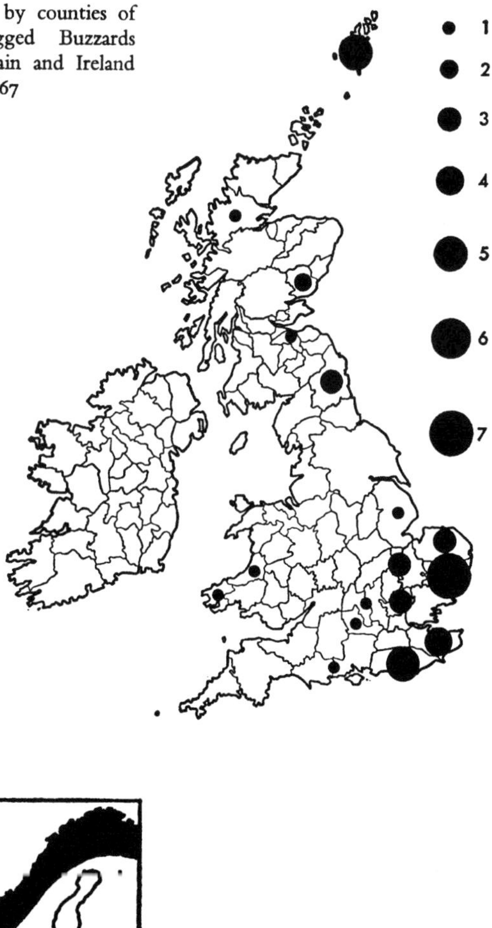

Fig. 38. European distribution of Rough-legged Buzzards *Buteo lagopus* with the breeding range shown in black and the normal wintering area enclosed by a dotted line (reproduced, by permission, from the 1966 edition of the *Field Guide*)

years combined (Fig. 36a). The distribution of these 43 is given in Fig. 37.

It is well known that the numbers of Rough-legged Buzzards in a breeding area fluctuate according to the population levels of Lemmings *Lemmus lemmus* and other Arctic rodents (see Curry-Lindahl 1961). Scott (1968) was able to show that the exceptional influx in Britain in 1966/67 was probably associated with a scarcity of small mammals in northern latitudes in 1966 and a consequent southwards and westwards shift in the centre of the Fenno-Scandian breeding range (Fig. 38), from which area he postulated that the British birds originated. The 1960/61 influx coincided with a peak in the Swedish Lemming population in 1960 and 1961—though, except in a few areas, the Swedish Rough-legged Buzzard population was apparently not in phase (Curry-Lindahl 1961). The presence of wintering Rough-legged Buzzards in Britain is probably also influenced by food availability. Their main diet here appears to be Rabbits *Oryctolagus cuniculus*, but Baxter and Rintoul (1953) suggested a correlation in Scotland between wintering numbers and vole plague years, this also influencing the distribution within Scotland.

The geographical distribution during the ten years within Britain and Ireland (Fig. 39) is strongly biased by the large numbers in the single winter of 1966/67, when there was a higher proportion than usual in south-east England. Nevertheless, it clearly shows the concentration in Suffolk, Norfolk, and Lincolnshire, these three counties accounting for no less than 39% of all the records in the ten years. One suspects, however, that there may have been more Scottish records than those shown, for in some areas the species is regarded as regular and occurrences are perhaps not always reported. On the other hand, Rough-legged Buzzards are exceedingly rare birds in Ireland, with only 26 records altogether up to the end of 1965 (Ruttledge 1966) and only one in the ten years. This is rather surprising, for there were a few records in most counties in the English Midlands, south-west England and Wales.

New arrivals appear to be so clouded by individuals wandering within Britain that it is difficult to determine a clear pattern (Fig. 40). The earliest ones, however, were confined to England north of a line from the Thames estuary to the Bristol Channel. New records were most widespread and numerous in October, November and early December, the only region with none in this period being the north of Ireland. The most static time was 10th December to 14th January, when there were reports of 'new' birds (presumably wandering wintering individuals) only from the main regions of wintering. From mid-January to late

March such new sightings became rather more widespread again (perhaps due to winterers wandering more widely as food supplies were reduced), but in late March and April the majority were in East Anglia and south-east England; these possibly represented a spring passage.

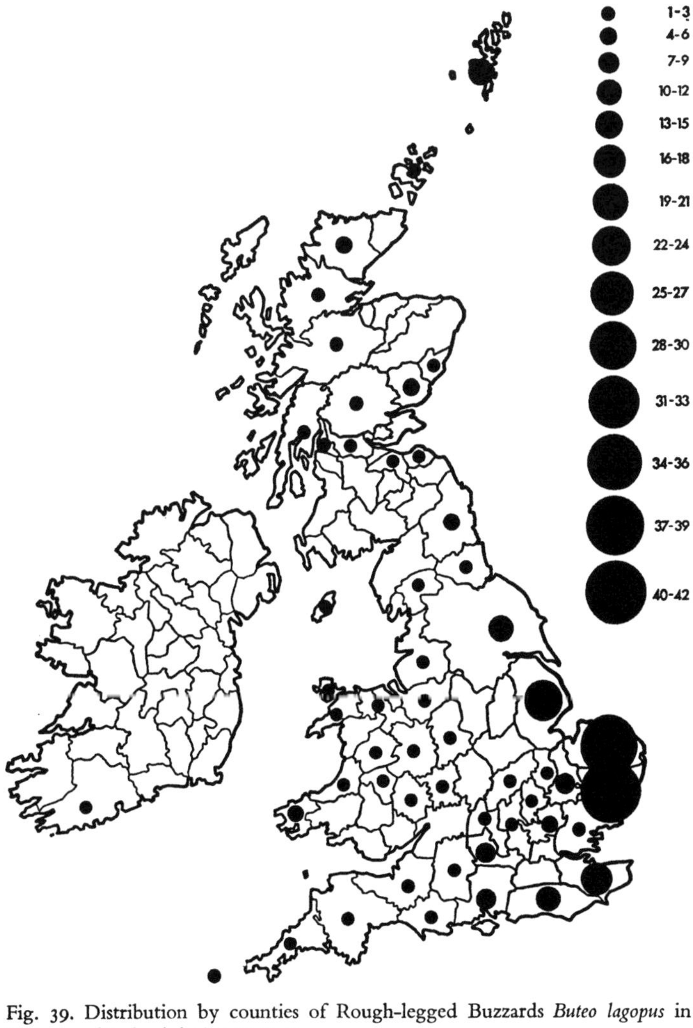

Fig. 39. Distribution by counties of Rough-legged Buzzards *Buteo lagopus* in Britain and Ireland during 1958–67

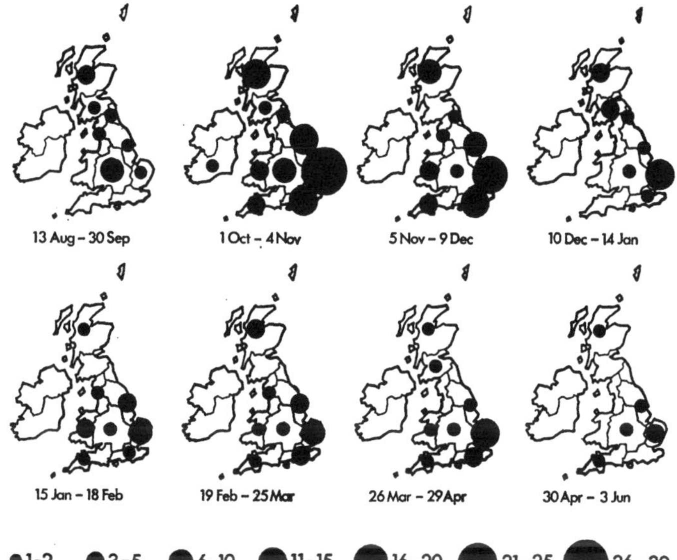

Fig. 40. Regional distribution in eight periods from autumn through to spring of Rough-legged Buzzards *Buteo lagopus* in Britain and Ireland during 1958–67

Temminck's Stint
Calidris temminckii

Considering that Temminck's Stints breed commonly as close to Britain as Scandinavia south to a latitude level with Shetland (Fig. 41), and nested successfully in Scotland in 1971 (Headlam 1972, Dennis 1972), they are remarkably scarce as migrants in Britain and Ireland. A total of only 291 was reported during the ten years, an average of 29 per year, which is only five more than the average of the commonest Nearctic vagrant, Pectoral Sandpiper. Records were most frequent in autumn, 64% compared with 36% in spring (Fig. 42). The spring passage extended from late April to the end of June, but 91% were in May, with the peak during 14th–20th. The autumn passage was much more protracted, extending from mid-July to mid-November, with the peak during 3rd–9th September. The extended passage in autumn is shown by only 59% occurring during the peak five weeks (20th August–23rd September), compared with the 91% during the peak five weeks in spring.

The spring records were mainly in the eastern half of England, over 40% of them in Norfolk, Kent and Suffolk, but there was still a relatively large proportion (17%) inland in the Midlands (Fig. 43). Hoopoes are well known as scarce migrants inland, yet only 11% of the spring records of that species were in the Midlands (Table 1). The autumn distribution of records was similar to that in spring, but, while there

Fig. 41. European distribution of Temminck's Stints *Calidris temminckii* with the breeding range of this summer visitor shown in black (reproduced, by permission, from the 1966 edition of the *Field Guide*)

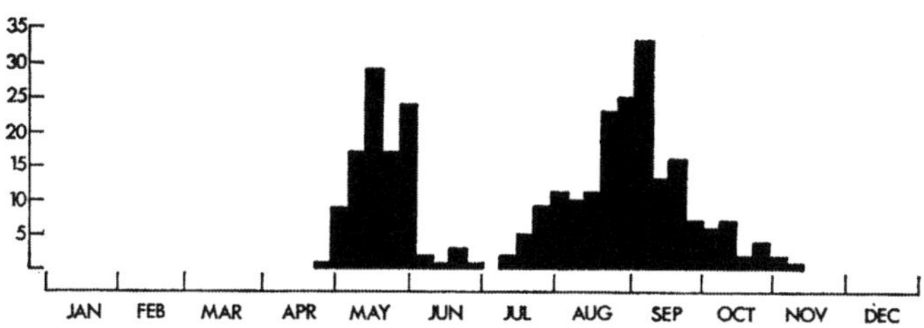

Fig. 42. Seasonal pattern of Temminck's Stints *Calidris temminckii* in Britain and Ireland during 1958–67

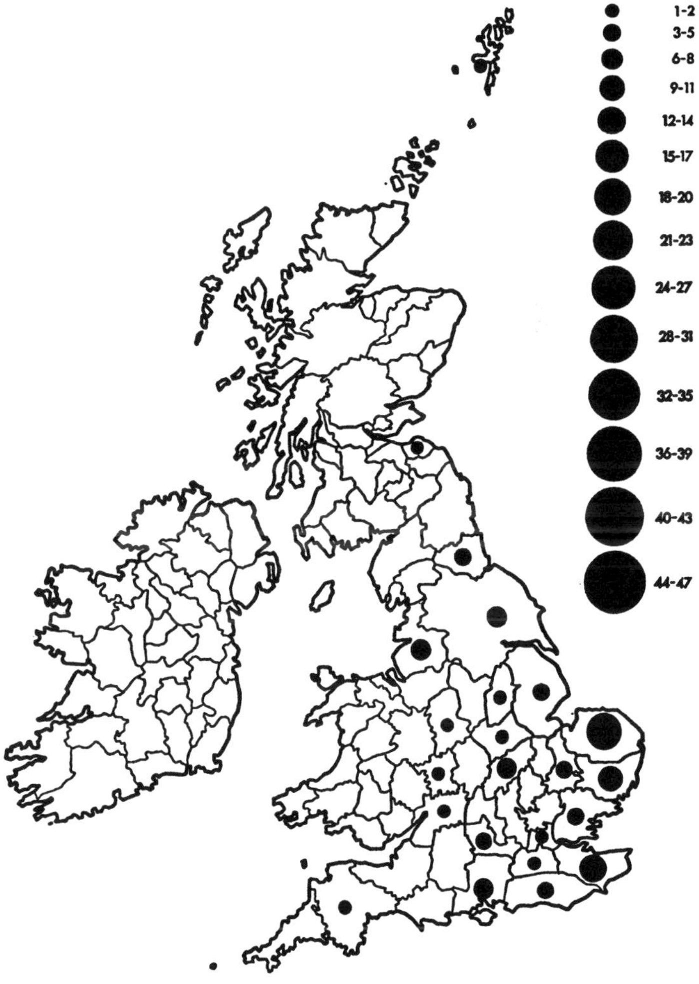

Fig. 43. Distribution by counties of spring Temminck's Stints *Calidris temminckii* in Britain and Ireland during 1958–67

was an even higher proportion (almost 50%) in the three main counties of Kent, Suffolk and Norfolk, the others were more widespread, with a few in Ireland and Wales (where there was none in spring) and more in south-west England (Fig. 44).

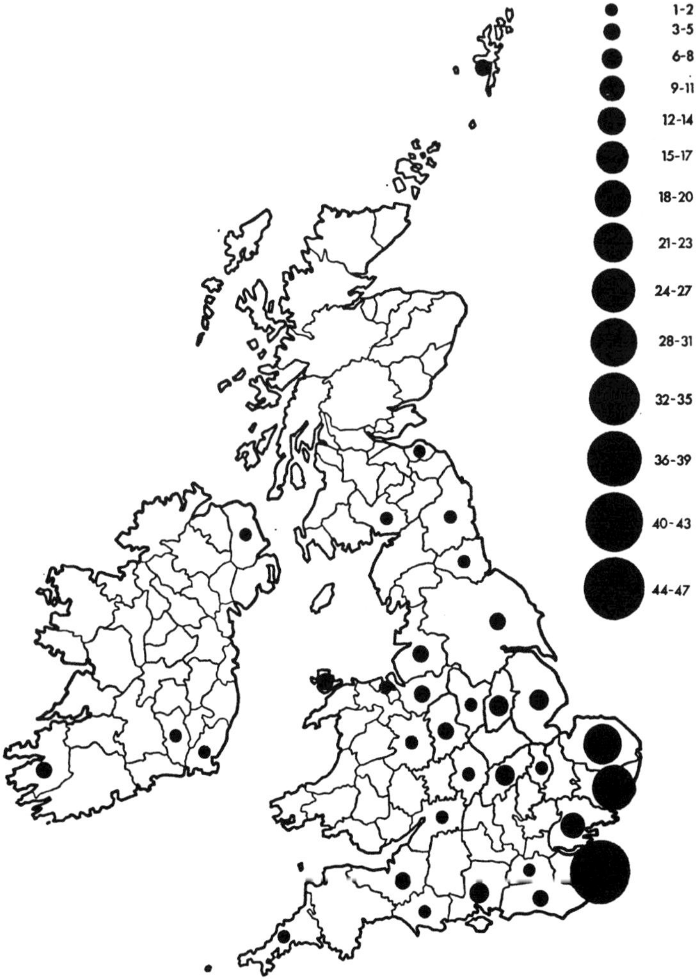

Fig. 44. Distribution by counties of autumn Temminck's Stints *Calidris temminckii* in Britain and Ireland during 1958–67

The numbers occurring annually in Britain and Ireland varied from two to 16 in spring (with peaks in 1966, 1964 and 1961) and from ten to 26 in autumn (with peaks in 1961, 1965 and 1963) (Fig. 45). These annual numbers may be compared with those of some other northern waders. The Little Stint *C. minutus* has a more northerly and easterly

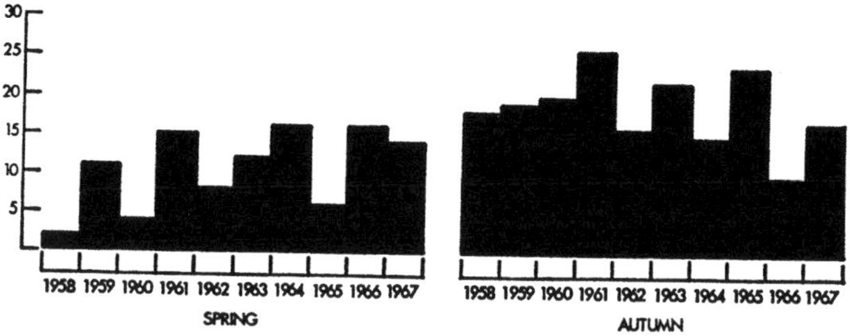

Fig. 45. Annual pattern of Temminck's Stints *Calidris temminckii* in Britain and Ireland during 1958–67 with the spring and autumn records shown separately

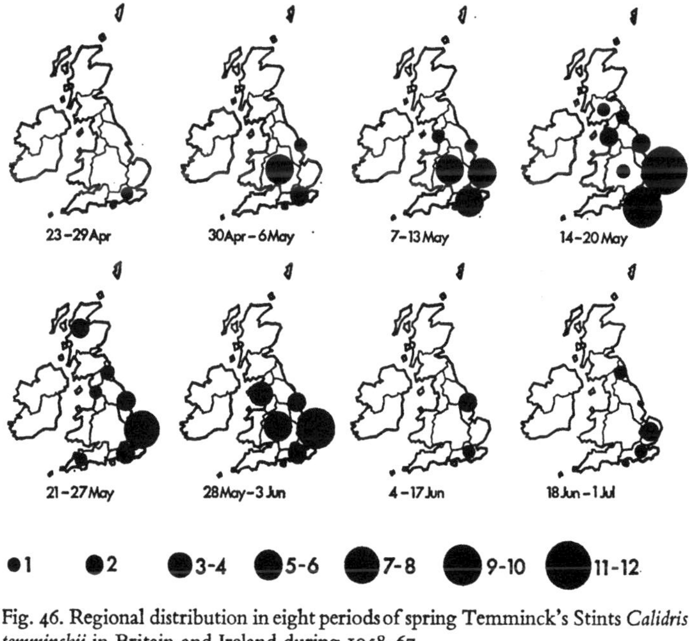

Fig. 46. Regional distribution in eight periods of spring Temminck's Stints *Calidris temminckii* in Britain and Ireland during 1958–67

breeding distribution than Temminck's, yet occurs here in autumn in far larger numbers. The Broad-billed Sandpiper *Limicola falcinellus* has a Scandinavian breeding distribution similar to that of Temminck's Stint, yet is much rarer in Britain. The breeding distribution of the Curlew Sandpiper *C. ferruginea* is well to the east, in arctic Asia, yet

Table 4. Records of four northern waders—Temminck's Stint *Calidris temminckii*, **Broad-billed Sandpiper** *Limicola falcinellus*, **Little Stint** *C. minutus* **and Curlew Sandpiper** *C. ferruginea*—**in Britain and Ireland during 1958–67**

The assessments of the last two species are taken from 'Recent reports' in the journal *British Birds*; whilst not pretending to be complete, this is the only readily available source of such information

	SPRING		AUTUMN			
	Temminck's Stint	Broad-billed	Temminck's Stint	Little Stint	Broad-billed	Curlew Sandpiper
1958	2	0	18	'normal'	0	'extremely scarce'
1959	11	1	19	'good numbers'	0	'avalanche'
1960	4	0	20	'unprecedented'	2	'possibly above average'
1961	15	0	26	'fairly numerous'	2	'below average'
1962	8	1	16	'rather scarce'	1	'rather scarce'
1963	12	2	22	'high but less than 1960'	1	'above average'
1964	16	0	15	'possibly above average'	0	'possibly above average'
1965	6	0	24	'at least one big influx'	0	'at least one big influx'
1966	16	0	10	'not particularly numerous'	0	'not particularly numerous'
1967	14	0	17	'large passage'	2	'large passage'

occurs on autumn passage in numbers similar to those of the Little Stint. The numbers of the two rarer species and rough assessments of the two commoner species are given in Table 4. There is virtually no relationship between the records of Temminck's Stint and the other three species (though Curlew Sandpipers and Little Stints do appear to show some degree of correlation). Even though rarer in Britain, the numbers of Temminck's Stints fluctuate far less than those of Little Stints and Curlew Sandpipers. This suggests that the rarer species is a regular migrant with meteorological conditions not of prime importance in its occurrence and also that it may be less subject to violent population fluctuations than are the other two.

The earliest spring arrivals were in south-east England and, particularly, the Midlands in the first week of May (Fig. 46). By the second week of May there were equal numbers in south-east England, East Anglia and the Midlands. A week later, the time of the main peak, the majority were in East Anglia and south-east England, with scattered records north to southern Scotland. During the last two weeks of the peak passage most were in East Anglia and the Midlands, with the pro-

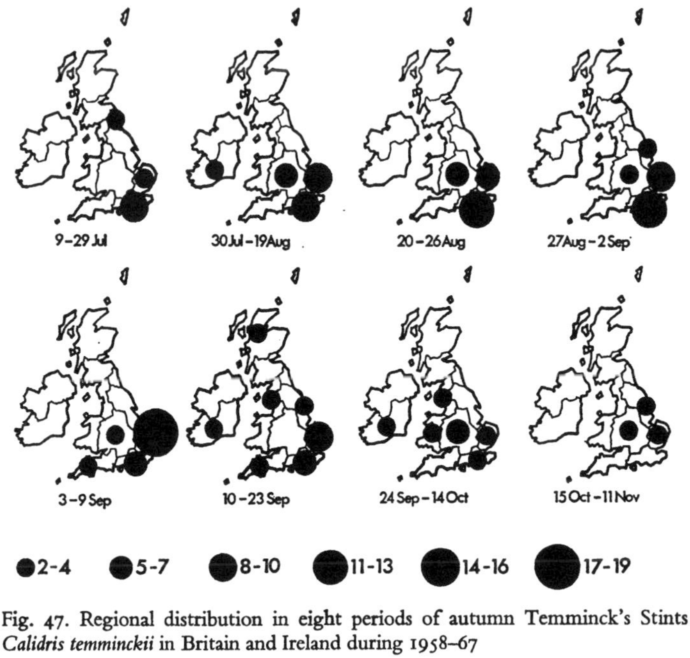

Fig. 47. Regional distribution in eight periods of autumn Temminck's Stints *Calidris temminckii* in Britain and Ireland during 1958–67

portion in south-east England dropping. The pattern suggests that the early arrivals may be mainly genuine passage migrants (presumably the westernmost part of the population) on their way north, whilst the later ones may include some which have been displaced westwards from more easterly areas, as well as a continuation of those on normal passage. Certainly the occurrences at the same localities on similar dates for two, three or even four successive years, preceded and followed by several years without any records there, suggests that the same individuals may be involved and confirms the proposition of a genuine passage of birds following set routes.

The long autumn passage shows a less clear pattern than that in spring. The only features of note are that the inland records in the Midlands made up a smaller proportion than in spring, but reached their peak earlier (20th–26th August) than the main concentration in East Anglia (3rd–9th September) (Fig. 47). As with the spring records, this tends to suggest that those in the Midlands were genuine passage migrants originating from the more westerly portions of the population, whilst those later in East Anglia were from more easterly and northerly portions. The situation is confused, however, for the peak in south-east England (20th August–2nd September) preceded that in East Anglia and, due to the relatively small numbers involved, the apparent differences may not be significant. The relative numbers in Norfolk and Kent in spring and autumn (more in Norfolk in spring and more in Kent in autumn) were also the reverse of what might logically be expected and one really requires more data before drawing firm conclusions.

Long-tailed Skua
Stercorarius longicaudus

The Long-tailed Skua is by far the rarest of the four skuas in Britain and Ireland. There was a total of only 171 records in the ten years and this species is thus the rarest of the 15 scarce migrants which are included in this book but which are not considered by the Rarities Committee.

Field-separation from small, long-tailed Arctic Skuas *S. parasiticus* is not easy, however, and some may go undetected (on the other hand, some 'Long-tailed Skuas' may have been misidentified Arctics). Immatures are even more difficult and it is probable that most of the 171 Long-tailed Skuas were adults, though the age was published in only 51 records (35 adults and 16 immatures).

All but 17 of the records were in autumn, from July to early Novem-

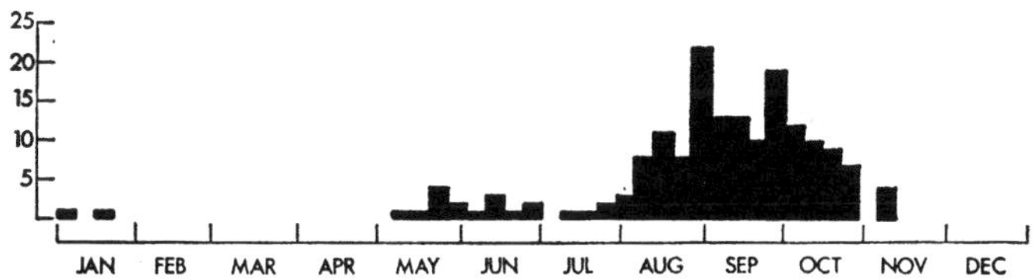

Fig. 48. Seasonal pattern of Long-tailed Skuas *Stercorarius longicaudus* in Britain and Ireland during 1958–67

ber (mainly August–October), with marked peaks at the end of August and the end of September (Fig. 48). There were 15 in May or June and two in January. The 15 spring records were mainly in two areas, south-western Britain (five) and northern Scotland (six), the other four being scattered along the British east coast (Fig. 50). The much more numerous autumn records were mainly on the English east coast, with almost four times as many in Norfolk as in any other county and over two-thirds of the total in the four contiguous counties of Durham, Yorkshire, Lincolnshire and Norfolk (Fig. 49).

The Long-tailed Skua has an arctic breeding distribution, mainly between the 33° and 59°F July isotherms (Voous 1960), with the westernmost part of the population as close to Britain as southern Norway (Fig. 51). There is known to be a close relationship with the roughly four-year cycle of Lemmings and other northern rodents, but the British and Irish records in the ten years do not appear to reflect this at all, the autumn totals varying from seven to 24 and the peak years being 1959, 1961, 1963, 1966 and 1967 (Fig. 52). Neither is there any apparent correlation with the numbers of Rough-legged Buzzards (which have a very similar breeding distribution and also fluctuate with the Lemming population), for these reached peaks in the autumns of 1960, 1962 and 1966 (Fig. 36b).

The wintering area of the Long-tailed Skua is still not known for certain, but is thought to lie far to the south, possibly off South America (Bourne 1967), making the two January records (Anglesey in 1963 and

Norfolk in 1967) highly remarkable. Ship observations have shown (1) that, whereas the passage of Great Skuas *S. skua* is relatively evenly spread across the Atlantic, the other three European skuas—Arctic, Pomarine *S. pomarinus* and Long-tailed—are more concentrated between 30° and 50°W, with the last-named the most westerly of all in

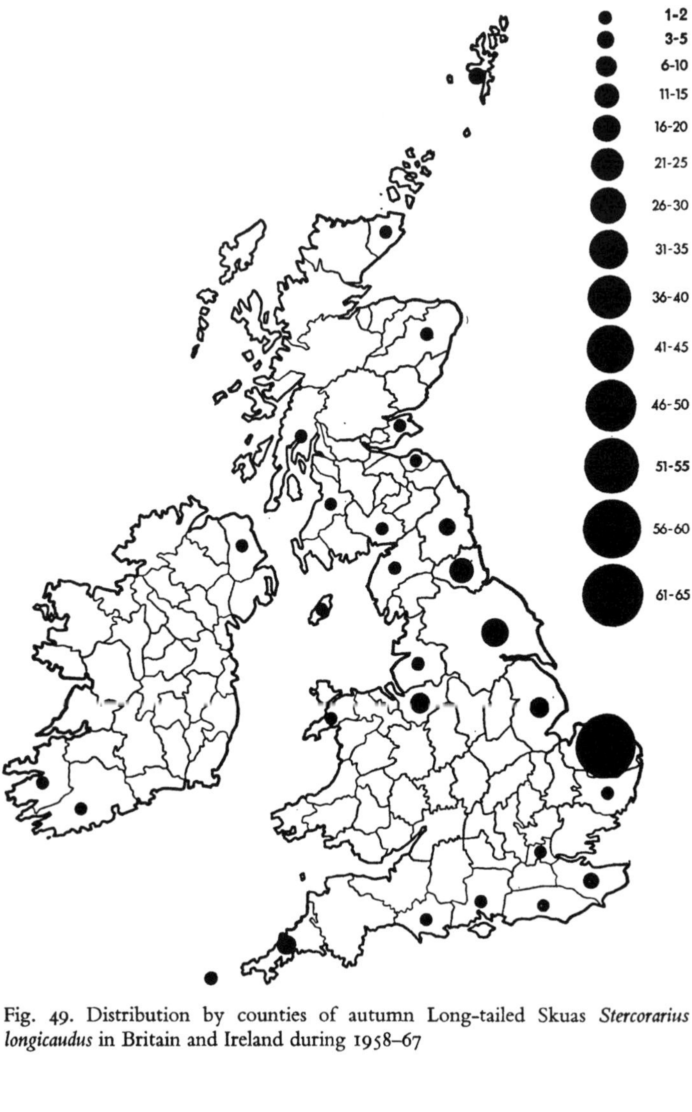

Fig. 49. Distribution by counties of autumn Long-tailed Skuas *Stercorarius longicaudus* in Britain and Ireland during 1958–67

Fig. 50. Distribution by counties of spring Long-tailed Skuas *Stercorarius longicaudus* in Britain and Ireland during 1958–67

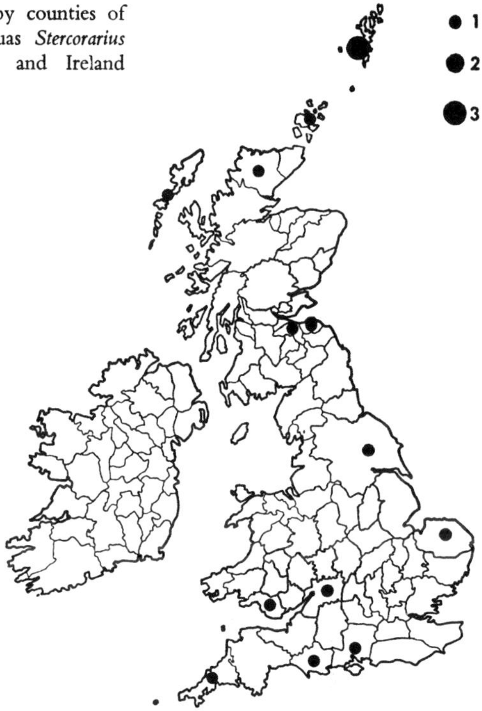

both spring and autumn (Tuck 1966); (2) that Long-tailed Skuas are the least often seen, but when they do occur they are often in flocks of up to 50 or more and sometimes as many as 160 have been recorded in a single day (Aikman 1966); and (3) that autumn passage seems to be quite prolonged, from July well into September (Sage 1968), whereas spring passage in the north Atlantic is mostly in May (with fewer in April and June).

Many of the observations in the western North Atlantic undoubtedly refer to birds from arctic Canada, but most authors have assumed that the Old World populations follow a similar west Atlantic route. For instance, Bell (1965) postulated that the European population passes between Iceland and the Hebrides, far out to sea, to join birds from Greenland and eastern North America in following a common route southwards in autumn; and that the reverse route is taken in spring, the Finnish population arriving overland from the west across the

Fig. 51. European distribution of Long-tailed Skuas *Stercorarius longicaudus* with the breeding range of this summer visitor shown in black (reproduced, by permission, from the 1966 edition of the *Field Guide*)

north Norwegian mountains and not via the Baltic. He and others have also drawn attention to the relatively high incidence of inland records in Europe and have suggested that this is a reflection of the weakness of this species, compared with the other skuas, in stormy conditions, though it might only reflect the species' use of overland routes on occasions. The largest single authenticated 'wreck' of Long-tailed Skuas ever recorded in British waters occurred on the Yorkshire coast on 7th–19th October 1879 (Nelson 1907), but the grand total for that period was only about 30 individuals, compared with 5,000–6,000 Pomarine Skuas at the same time.

The great preponderance of Norfolk records (Fig. 49) might be the result of displacement southwards by northerly gales, but, if so, it poses a problem: if the species is thus prone to displacement in stormy conditions, it is surprising that so few have been recorded in autumn in northern Scotland and the north of Ireland. In this connection, the observatories at Malin Head and Tory Island, Co. Donegal, and The Mullet, Co. Mayo, are primarily sea-watching stations, yet none has produced a single record. Some time ago it was pointed out that the violent westerly weather of September 1950 did not result in any Long-tailed Skuas being recorded in Britain or Ireland (Anon 1952). Further, 1,628 hours of systematic timed sea-watching was carried out in the months August–October at Cape Clear Island, Co. Cork, in 1959–67 and totals of 2,106 Great, 292 Arctic and 54 Pomarine Skuas were recorded, but not one Long-tailed, despite frequent westerly or south-westerly gales bringing spectacular sea-passage of other species. The only conclusion must be that they are far commoner in autumn in the North Sea than they are in the North Atlantic, especially considering the occurrence in Britain and Ireland of over 500 Nearctic waders during the ten years (about three times as many as the total of Long-tailed Skuas in the same period). If Long-tailed Skuas were really such weak fliers, one would expect many more to be recorded on the western coasts. There is no longer the excuse that these areas are under-watched, for due to the establishment of the observatories already mentioned and the activities of the Seabird Group's 'Atlantic Sea-watch Scheme', there has probably been more systematic sea-watching carried out in western Ireland than anywhere else in Britain and Ireland.

There appears to be little or no evidence for any passage of Long-tailed Skuas through the English Channel: sea-watching at Dungeness, Kent, produced only one record in 1958–67; there was also only one at Cap Gris Nez, France, in the three years 1965–67; and very few are recorded in the Biscay area or off north-west Spain (e.g. Pettitt 1969), where the other three species are seen in large numbers. It seems unlikely, therefore, that there is any substantial passage either through the English Channel or overland from Scandinavia across south-eastern England to Finisterre. This makes the protracted period of occurrences and relatively large numbers in autumn in the southern North Sea (on the north Norfolk coast) rather perplexing, unless it is postulated that, after the end of the breeding season, part at least of the population is present for about two months feeding in the northern North Sea. Bell (1965) suggested that those seen on the English and Scottish east coasts in autumn were likely to be Fenno-Scandian in origin. The extended period over which records occur, together with their relative absence from the other north-facing parts of Britain and Ireland,

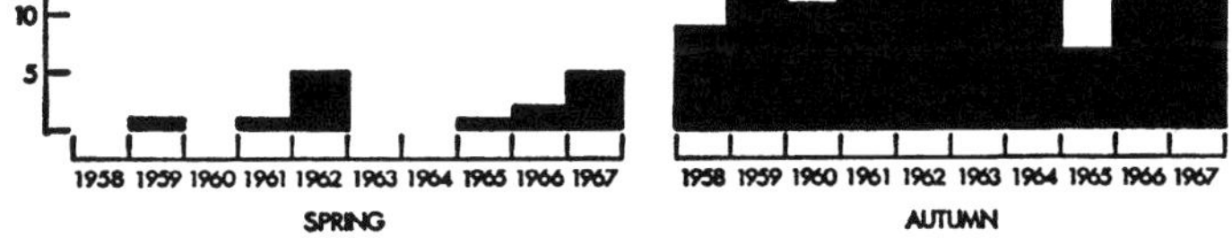

Fig. 52. Annual pattern of Long-tailed Skuas *Stercorarius longicaudus* in Britain and Ireland during 1958–67 with the spring and autumn records shown separately

indicates, however, that they may involve individuals which stay feeding in the area rather than ones truly on passage and that, therefore, there is every reason to suspect that they may include some from further east, in Asiatic Russia. The likeliest situation is that migrant Long-tailed Skuas delay for a considerable time in the rich feeding grounds of the northern North Sea, but that the population there is always changing, originating from further and further east as the autumn progresses. Westward departures from the northern North Sea may take place only in ideal calm conditions and be a very rapid process. This would help to explain the paucity of sightings from northern and western Scotland and Ireland.

The distribution of records through the autumn (Fig. 53) reveals a pattern which it is tempting to interpret as reflecting an increasingly more southerly distribution in the North Sea from 20th August to 30th September (with fewer and fewer occurring in north-east and eastern England, but more and more in East Anglia) and then a northwards withdrawal during October (with fewer and fewer in East Anglia, but

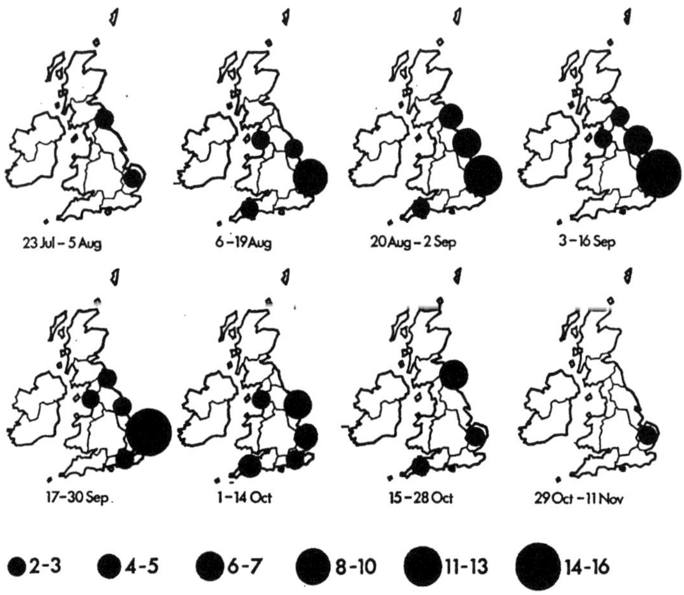

Fig. 53. Regional distribution in eight 14-day periods of autumn Long-tailed Skuas *Stercorarius longicaudus* in Britain and Ireland during 1958–67

second peaks of the autumn in, first, eastern England and, second, north-east England). Such small numbers are involved, however, that this attractive pattern may be entirely fortuitous. It has already been questioned whether the species is really such a weak flier compared with the other skuas. The records for the ten years seem rather to suggest that it occurs in the southern North Sea when the autumnal feeding grounds to the north are disturbed by rough weather and that the birds are not necessarily gale-blown, but may be positively seeking calmer areas for feeding (the species seems to be less piratical in its feeding habits than the other skuas). 'Wrecks' such as that of 1879 probably occur only after stormy conditions have persisted for a considerable time, depriving the birds of food and weakening them.

The dearth of spring records on western coasts also suggests that they are not easily storm-driven. The only spring 'wreck' on record (60–70 in the Shannon estuary on 16th May 1860) is excluded from the latest work on the birds of Ireland (Ruttledge 1966) as there is no proof of correct identification. Certainly, normal westerly gales in May do not bring this species to western Ireland, for in 116 hours of systematic sea-watching at Cape Clear Island in May 1962–67 91 Great, 78 Arctic and 51 Pomarine Skuas were recorded, but no Long-tailed. It is noteworthy that Pomarine Skuas are much commoner in spring than autumn at this station—peaks of 0·59 per hour in the first half of May and 0·08 per hour in the first half of October, based on 57 and 216 hours respectively (Sharrock 1973). These spring Pomarine Skuas occur, usually in small flocks, after or during gales and are clearly on their way north to the Arctic breeding grounds which are virtually identical to those of the Long-tailed Skua, with the exception that they do not extend westwards into Fenno-Scandia. If Long-tailed Skuas were so prone to storm-displacements as they are reputed to be, one would expect some occasionally to occur coincidentally with these Pomarine Skuas, even though they are found further west in the Atlantic. *The Handbook* and Bell (1965) noted that over half of the few spring records have been in north-western Ireland; this was not the case in 1958–67, when not one of the 17 spring records was in Ireland.

Finally, attention should be drawn to the apparently increasing frequency with which single Long-tailed Skuas are recorded in Scottish Arctic Skua colonies in summer, at a time when a number of other northern species are colonising northern Britain.

CHAPTER 4

Bluethroat and Ortolan Bunting

Both of these species are fairly widespread in Europe, with breeding ranges extending from Scandinavia to Iberia (Figs. 54 and 60). The Bluethroat is of particular interest, for the breeding males can be sub-specifically identified in the field, with a Scandinavian red-spotted race *L. s. svecica* and a central and southern European white-spotted race *L. s. cyanecula*.

Bluethroats are closely related to Robins *Erithacus rubecula* and Nightingales *Luscinia megarhynchos* and are Robin-sized. They are brown above and in all plumages have rufous patches at the base of the tail (the best single field-character) and a long creamy supercilium over the eye. In summer the males have bright blue throats with a red (*svecica*) or white (*cyanecula*) central spot, and a red lower border. Females and young birds lack the blue and red on the breast, but have a brown gorget. Bluethroats are often very difficult to see, diving into cover as soon as they are flushed (perhaps momentarily perching on a briar or twig before disappearing), but a patient watcher may see them out in the open, when their long legs, drooped wings and cocked tail give them a charming and distinctive silhouette. In their breeding areas they occur in marshy thickets but on passage are found mainly on the coast, in any area of tangled vegetation.

Ortolan Buntings are the same size as Yellowhammers *Emberiza citrinella* and adults are rather similar but lack the rufous of the Yellowhammer's rump and have an olive-green head and breast, yellow throat and orange-brown underparts. Immature Ortolan Buntings are brown and streaked, the best field-characters being a pink bill, pale eye-ring and distinctive liquid call-note, 'plwilk' or 'plwik'. They may be rather shy and difficult to approach and are frequently identified, as they fly over, from the diagnostic call-note alone, by observers familiar with it. Birds of scrub and gardens in their breeding areas, they occur on passage mostly in coastal areas, often with other buntings, finches and larks in stubble or ploughed fields.

Bluethroat
Luscinia svecica

A total of at least 600 Bluethroats was recorded in Britain and Ireland during the ten years and, although this is less than half the number of Hoopoes, it is thus the second most common species dealt with in this book. In fact, 600 is likely to be something of an under-estimate for this skulking bird. For instance, the county and regional reports documented about 122 individuals in the entire autumn of 1965, yet Davis (1966) considered that probably not many fewer than 280 were involved in the great immigration of early September 1965.

The sex and/or age of relatively few of the 600 individuals were published, but included 67 adults, 50 immatures, 102 males and 39 females. Further, only 99 of the 600 birds were subspecifically identified, with 66 as *svecica* and 33 as *cyanecula*. This proportion of 2:1 is probably an underestimate, however, since *cyanecula* is regarded as a rarity and every one identified is likely to be mentioned in detail in a local report, whereas *svecica* is generally recognised as regular and those identified may not, unfortunately, always be specified as such. (*The Handbook*

Fig. 54. European distribution of Bluethroat *Luscinia svecica* with the breeding range of this summer visitor shown in black (reproduced, by permission, from the 1966 edition of the *Field Guide*)

regarded *svecica* as a regular autumn passage migrant, but gave only eleven records of *cyanecula* which are now acceptable, following the investigations of Nicholson and Ferguson-Lees 1962.)

Four-fifths of the records were in autumn, from mid-August to mid-November, with a peak in early September (Fig. 55); and almost one-fifth were in spring, from mid-March to mid-June, with two peaks, one in late March and the other in late May. Outside these periods, there were single observations in January (Kent in 1960) and late June (Shetland in 1965). The records of each race were almost equally divided between spring and autumn. In spring, however, 16 of the 18 *cyanecula*, but only one of the 34 *svecica*, came before 23rd April. There was no such distinction in the autumn data. In spring, when sexing is usually (but not always) easy, there were 48 males, 30 females and 34 unsexed. In autumn, when birds other than adult males with remains of summer plumage are usually unsexable, there were 54 males, nine females and 423 unsexed.

More than half of the spring Bluethroats (57 out of 112) were recorded in Shetland, with small numbers in most of the English south and east coast counties (Fig. 56). The two races were unequally divided, however, with half of the *cyanecula* on the English south coast and nearly four-fifths of the *svecica* in Scotland (Table 5).

Table 5. Geographical distribution of spring Bluethroats *Luscinia svecica* in Britain and Ireland during 1958–67

	All records	White-spotted *L. s. cyanecula*	Red-spotted *L. s. svecica*
South-west England	8	3	1
South-east England	16	6	1
East Anglia	4	2	2
Midlands	1	1	0
Eastern England	6	2	0
North-west England	0	0	0
North-east England	6	1	3
Wales	0	0	0
South of Ireland	1	0	0
North of Ireland	0	0	0
Southern Scotland	6	0	3
Northern Scotland	64	3	24

The distribution of autumn records differed considerably from that in spring. Although substantial numbers again occurred in Shetland, these made up less than a fifth of the total (compared with over half

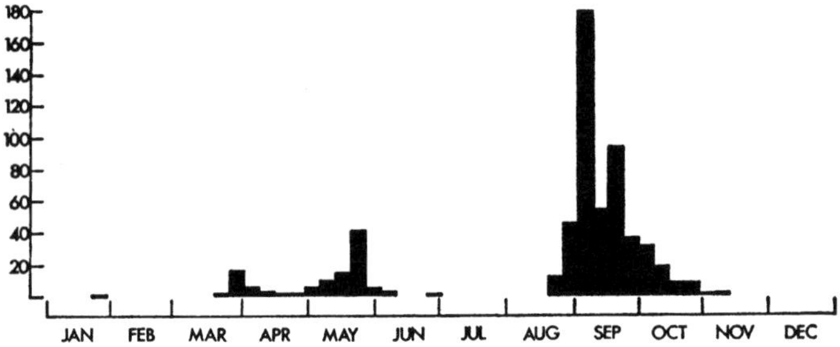

Fig. 55. Seasonal pattern of Bluethroats *Luscinia svecica* in Britain and Ireland during 1958–67

in the spring) and there was a much more even spread on the English east and south coasts (Fig. 57). The proportion in Suffolk was greatly influenced by the numbers occurring in the great immigration of 1965 (Davis 1966). The Bluethroat was not added to the Irish list until 1954, but even so the very low number recorded in Ireland, Wales and north-west England (only 15 in the ten years) is remarkable. The racial distribution was similar to that in the spring, with most of the *cyanecula* (twelve out of 14) on the south coast of England and four-fifths of the *svecica* (26 out of 32) in eastern England or northern Scotland (Table 6).

Table 6. Geographical distribution of autumn Bluethroats *Luscinia svecica* in Britain and Ireland during 1958–67

	All records	White-spotted *L. s. cyanecula*	Red-spotted *L. s. svecica*
South-west England	60	4	1
South-east England	79	8	2
East Anglia	139	0	0
Midlands	0	0	0
Eastern England	42	1	10
North-west England	2	0	0
North-east England	59	0	0
Wales	6	0	1
South of Ireland	6	0	0
North of Ireland	0	0	0
Southern Scotland	10	1	2
Northern Scotland	84	0	16

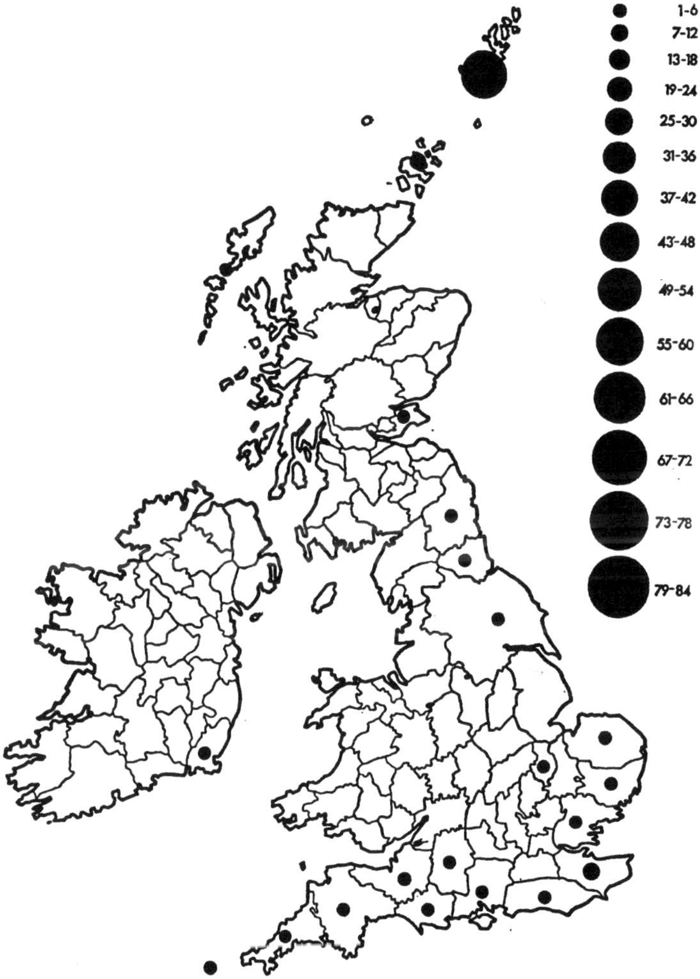

Fig. 56. Distribution by counties of spring Bluethroats *Luscinia svecica* in Britain and Ireland during 1958–67

The numbers of Bluethroats recorded each year varied from five to 25 in spring and from 19 to at least 122 in autumn (Fig. 58). As would be expected, the peak years for the two races did not coincide. In spring, half of the *cyanecula* occurred in 1958 and over half of the *svecica* in 1959 and 1960. Of the nine *cyanecula* in spring 1958, seven were recorded during the 14-day period from 19th March to 1st April, in Scilly, Hampshire, Sussex, Essex, Northumberland and Orkney, and this attracted attention at the time (Ferguson-Lees 1958). The proportion of autumn birds subspecifically identified is so small that conclusions cannot be drawn.

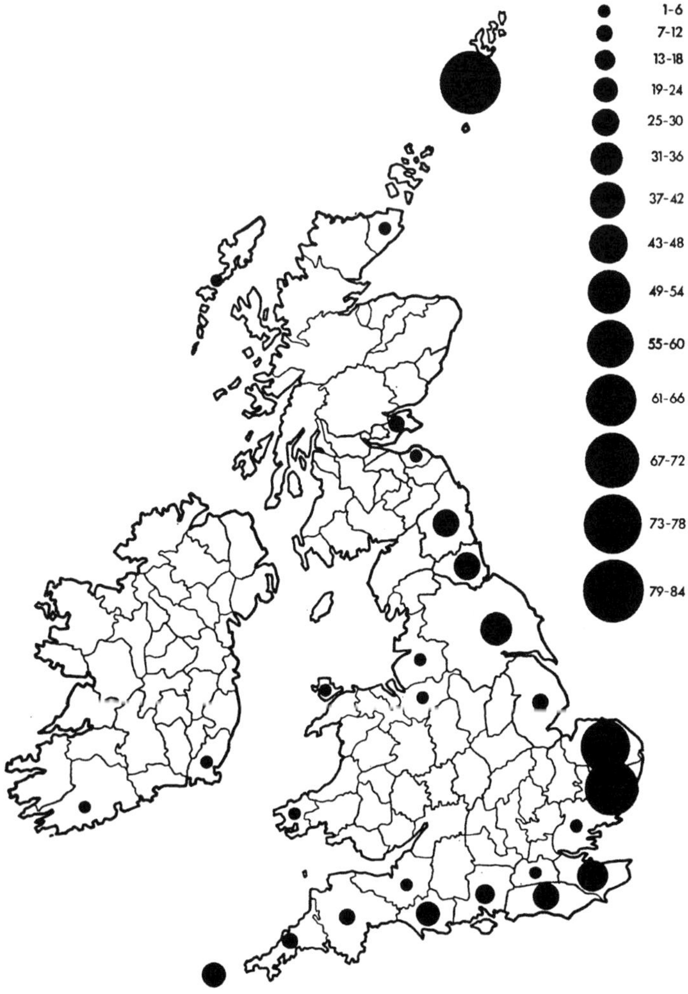

Fig. 57. Distribution by counties of autumn Bluethroats *Luscinia svecica* in Britain and Ireland during 1958–67

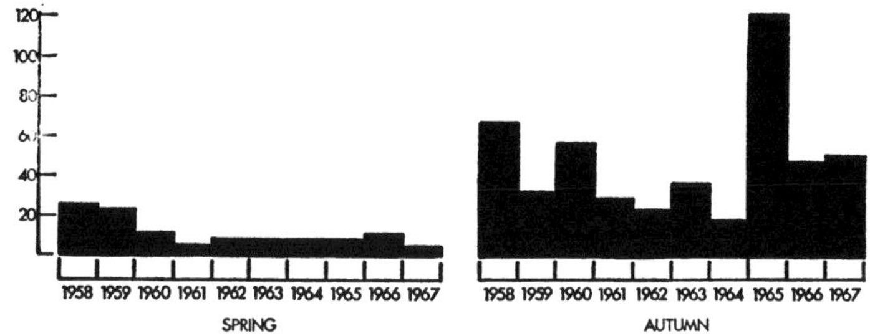

Fig. 58. Annual pattern of Bluethroats *Luscinia svecica* in Britain and Ireland during 1958–67 with the spring and autumn records shown separately

It is clear that Bluethroats are vagrants to Britain in spring. To a varying extent each year, small numbers of the south and central European white-spotted race *cyanecula* may overshoot on spring migration and occur, mainly on the English south coast, in late March or early April. Similarly, there is a rather greater likelihood of small numbers of the Scandinavian red-spotted race *svecica* overshooting and occurring, chiefly on the north Scottish islands, in late May.

Although the average number of Bluethroats recorded per autumn is only 49 (or 41 if the exceptional 1965 records are excluded), this is sufficient to justify their being regarded as regular migrants rather than vagrants. Lack (1960) showed that Bluethroats (unlike Red-breasted Flycatchers, Barred Warblers and Icterine Warblers) occurred on the east coast coincidentally with big arrivals of Continental migrants. It seems reasonable to assume that the westernmost SSW-oriented *svecica* regularly pass over or close to the English east coast in autumn (*cf.* Evans 1968) and that a varying proportion are grounded each year. The smaller number of *cyanecula* recorded mainly on the south coast at the same time pose a similar problem to south European species which occur north of their breeding ranges in autumn (e.g. Melodious Warblers in Ireland and Wales). The situation is not identical, however, for the Bluethroats probably include a higher proportion of adults (otherwise it would not be known that they were *cyanecula*) and the European breeding distribution is such that relatively little westerly displacement is necessary to explain vagrancy of *cyanecula* on the English south coast in autumn.

Outside the period under review, the first known case of breeding of Bluethroats in Britain was established in Scotland in 1968 (Greenwood 1968).

Ortolan Bunting

Emberiza hortulana

A total of 335 Ortolan Buntings was recorded in Britain and Ireland in the ten years. Published data on ages and sexes included 35 adults, 35 immatures, 46 males and 15 females. Many of the 265 not aged will probably have been immatures (adults, being noteworthy, are more likely to be mentioned in county reports), though others will have been birds heard but not seen closely. I consider the sex data to be very unreliable (unsexed adults not infrequently being called adult males).

About 15% of the records were from mid-April to June and 85% from mid-August to mid-November. Even when records for the ten years are combined (which tends to mask sharp peaks), the spring and autumn peaks were both very distinct, during 7th–13th May and 3rd–9th September (Fig. 59).

The European breeding distribution of the Ortolan Bunting (Fig.

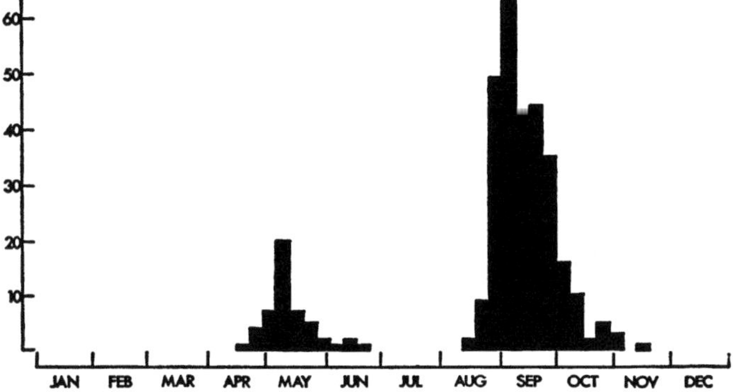

Fig. 59. Seasonal pattern of Ortolan Buntings *Emberiza hortulana* in Britain and Ireland during 1958–67

60) is similar to that of the Bluethroat (Fig. 54), though it extends further south into the Mediterranean countries. While the species is common in the south and also in Fenno-Scandia, it is, as noted by Nisbet (1957), less numerous in the intervening area. The pattern of spring vagrancy (Fig. 61), with three-fifths of the records in Shetland, is very similar to that of the Bluethroat (Fig. 56). The relative absence of records from the English south coast suggests that spring vagrancy results almost entirely from the overshooting (or diversion in excep-

Fig. 60. European distribution of Ortolan Bunting *Emberiza hortulana* with the breeding range of this summer visitor shown in black (reproduced, by permission, from the 1966 edition of the *Field Guide*)

tional weather conditions) of the Fenno-Scandian population. A notable example of this occurred outside the period under review, when 32 were seen on Fair Isle, Shetland, on 3rd May 1969 and the associated species, including 45 Wrynecks *Jynx torquilla*, 300 Ring Ouzels *Turdus torquatus* and 500 Bramblings *Fringilla montifringilla* (see Dennis 1970), clearly indicated the origin of the arrival.

There was not a great deal of difference in the timing of the spring movements from one region to another, but all of the six English south coast records and a third of the nine East Anglian ones were before 7th May, whereas nine-tenths of the 31 in Scotland were after 6th May. This is a similar pattern to that shown for several of the scarce migrants already dealt with (e.g. Hoopoe) and is what one would expect to find.

The distribution of autumn records (Fig. 62) was, like that of the Bluethroats, more widespread than in spring. In the ten years as many were recorded in Norfolk as in Shetland and these two counties with Scilly, Devon, Dorset, Pembrokeshire and Co. Cork accounted for almost 70% of the total. It is noteworthy that half of the autumn records were in western Britain and Ireland, compared with only 14% of those in spring. Before the period under review, the largest recorded influx of Ortolan Buntings occurred on the Isles of Scilly where at least 100 were seen on 25th September 1956.

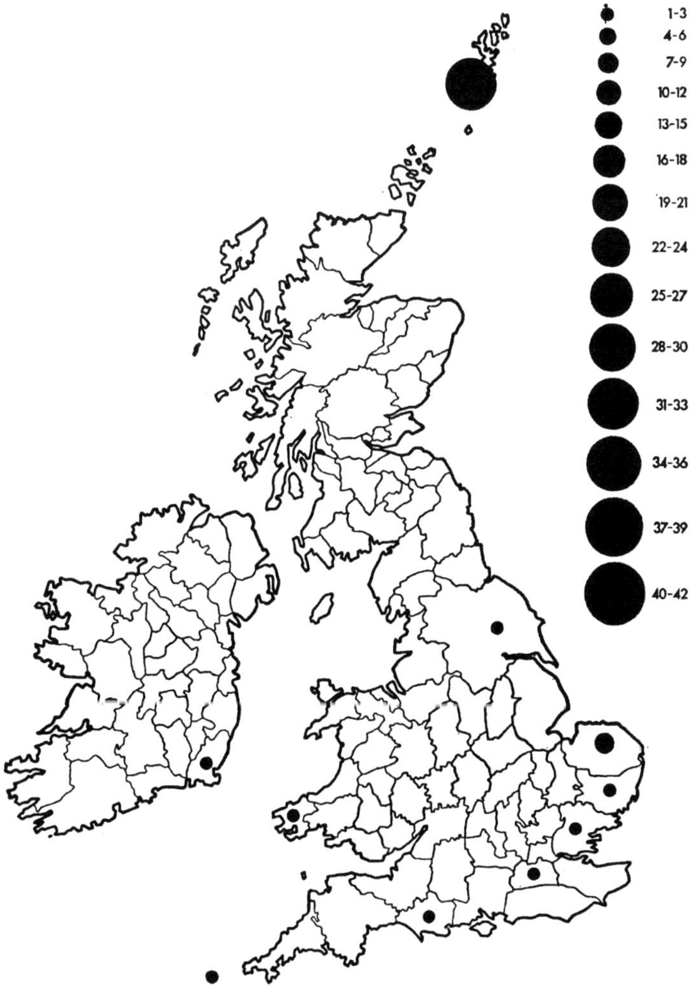

Fig. 61. Distribution by counties of spring Ortolan Buntings *Emberiza hortulana* in Britain and Ireland during 1958–67

As with several of the species already considered, the autumn records in south-western Britain and Ireland were substantially later than those in eastern Britain (Fig. 63). The first in mid-August were mainly in the Northern Isles, but by the end of the month there were substantial arrivals in East Anglia. Northern Scotland, East Anglia

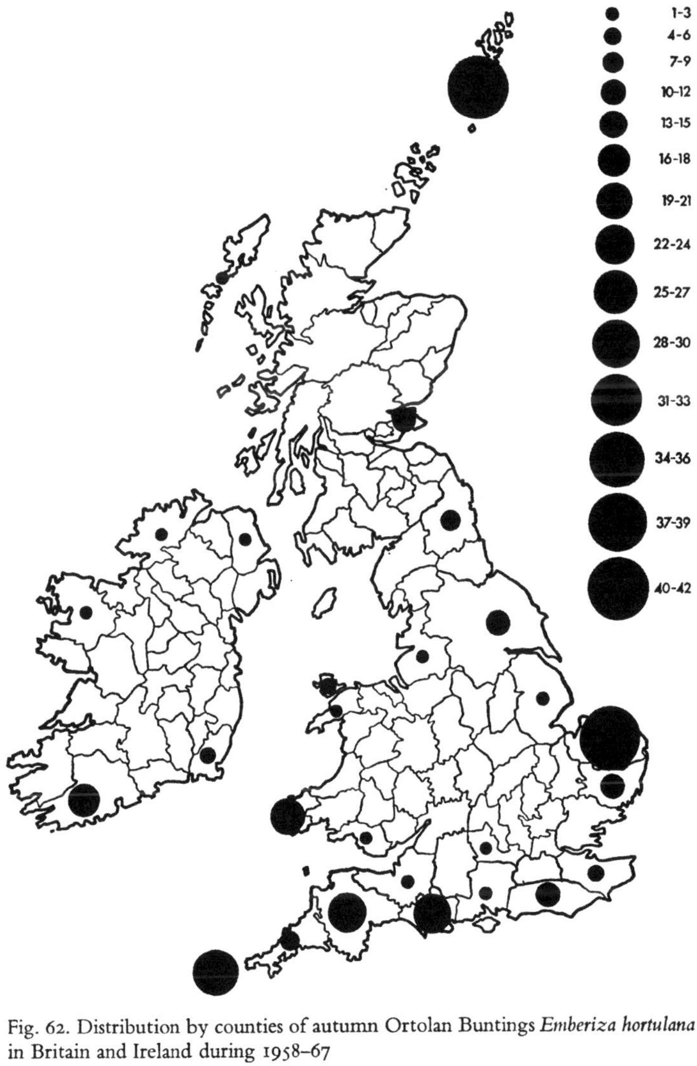

Fig. 62. Distribution by counties of autumn Ortolan Buntings *Emberiza hortulana* in Britain and Ireland during 1958–67

and south-west England all featured almost equally in early September, but from 10th September onwards there were always more in south or south-western areas than on the British east coast.

The numbers recorded annually in the ten years varied from one to ten in spring and from 19 to 43 in autumn (Fig. 64). Of the ten in spring 1967, nine were during 7th–13th May, eight of them in Shetland, but this influx paled into insignificance by comparison with that in 1969, already mentioned. The peak autumn numbers in 1965 coincided with those of Bluethroat and several other species, associated with the great immigration in early September (Davis 1966), but only eight of the 43 in that autumn were in East Anglia (compared with at least 81 of the 122 or more Bluethroats), so there was no direct meteorological connection. Williamson (1959) noted some 40–50 Ortolan Buntings in the September 'drift-movements' of 1956 and these, together with the 100 on one date in the Isles of Scilly, already mentioned, must make 1956 the peak year for the species' occurrence in Britain and Ireland.

It has at times been suggested (e.g. Lack 1961) that some of the

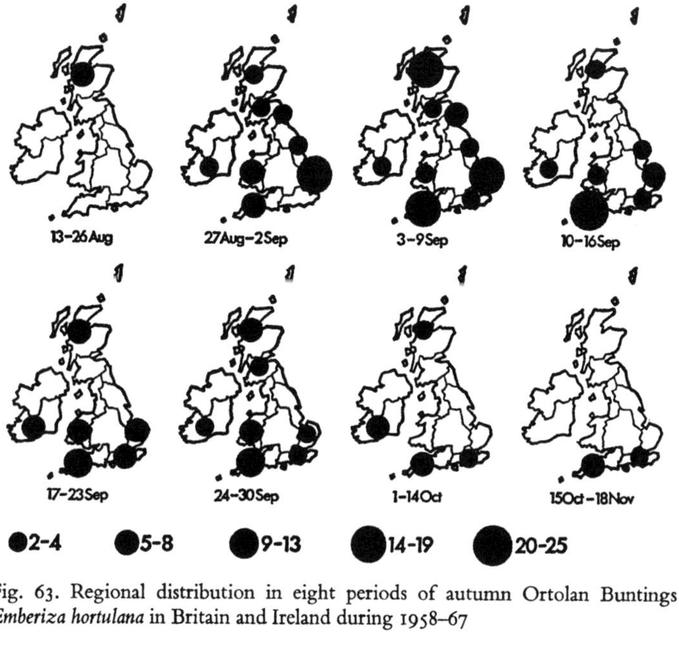

Fig. 63. Regional distribution in eight periods of autumn Ortolan Buntings *Emberiza hortulana* in Britain and Ireland during 1958–67

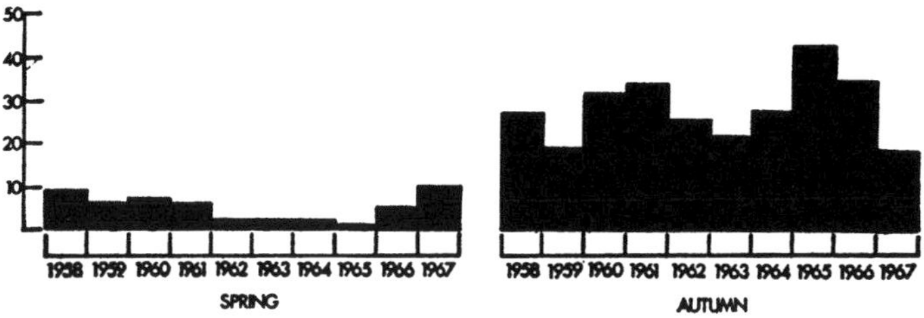

Fig. 64. Annual pattern of Ortolan Buntings *Emberiza hortulana* in Britain and Ireland during 1958–67 with the spring and autumn records shown separately

vagrants and other scarce species in western Britain and Ireland in autumn may be migrants grounded after diversion from a southerly or SSW course from Scandinavia to Iberia across Britain. Several features, however, seem to make this unlikely and the Ortolan Bunting picture is an example. Many species, Ortolan Bunting included, reach their peak a week or fortnight later in south-western Britain and Ireland than on the east coast. If the western birds resulted from grounded migrants overflying on a southerly course, one would expect a discrepancy of only a day or so, which would not appear in such crude analyses as these based on seven-day periods. Thus the records are consistently at variance with this hypothesis. Alternatively, it might be suggested that birds displaced from this stream, to northern Scotland, continue on a more westerly route southwards and result in the occurrences in western Britain and Ireland. Apart from meteorological objections, however, it must be noted that where Fenno-Scandian birds only are involved (e.g. Red-spotted Bluethroats) they are excessively scarce in western Britain and Ireland in autumn, suggesting that this situation does not appertain either. On the other hand, birds labelled as being of southern European origin (e.g. Melodious Warbler) are shown to occur in these western districts with regularity. Were it not for their labelling, they might well be attributed to Fenno-Scandian origin. It seems logical to assume that most of the Ortolan Buntings recorded in western Britain and Ireland in autumn (as many as on the east coast) similarly arrive from the south and either are derived from southern populations or are Scandinavian birds which have previously moved southwards through Continental Europe.

CHAPTER 5

American birds

In this chapter we turn to those species with a mainly North American, or Nearctic, breeding distribution. Two species, Pectoral Sandpiper and Sabine's Gull, outnumber all others, but we shall briefly consider for comparison the other Nearctic waders and gulls and also the Nearctic land-birds. The occurrences of Nearctic waterfowl in Europe have been summarised by Bruun (1971).

Pectoral Sandpipers are usually to be found in damp grass or sedge areas close to freshwater or brackish pools, when on this side of the Atlantic. They are between Dunlins *Calidris alpina* and Common Sandpipers *Tringa hypoleucos* in size; the sharp demarcation between the streaked upper breast and white lower breast and belly, and creamy 'V-shaped' lines on the back are the best field-characters.

Sabine's Gulls are usually seen in inshore waters only after gales. They have a strikingly contrasted wing-pattern, with black outer primaries, grey wing coverts and white inner primaries and secondaries (forming three distinct triangles), and this, together with a deeply forked tail, separates them from all other gulls.

The other species of waders and gulls, and the Nearctic land-birds, considered in this chapter are well-illustrated in field guides such as Robbins *et al.* (1966) and Peterson (1947).

Pectoral Sandpiper

Calidris melanotos

and other American waders

A total of about 62 Pectoral Sandpipers was recorded in Britain and Ireland up to 1940, less than 30 in the next decade and under 60 in the seven years 1951–57 (*The Handbook*, Anon 1952, Nisbet 1959, Williamson 1963). Individuals at nine localities in the autumn of 1948 were described at the time as 'a miniature invasion' (Anon 1949) and ten or fewer in 1950 were also considered 'sufficient to constitute an invasion' (Anon 1951). Yet a total of 243 was recorded during 1958–67, an average of about two dozen a year. It is impossible to know the extent to which this increase merely reflects the growing number of observers and their greater awareness of the possibility of finding Nearctic waders.

Of the 243 recorded in the ten years, no less than 235 (97%) were in autumn (July–early November) and only seven in spring (April–May); there was also one in December. The peak fell clearly in the middle fortnight of September and the distribution shows an interesting negative skewness (Fig. 65). The seven spring records were all in the west—single individuals seen in Co. Antrim, Co. Derry, Co. Kerry, Co. Cork, the Isles of Scilly, Cornwall and Pembrokeshire. The autumn records, however, were more widespread (Fig. 66), Co. Kerry

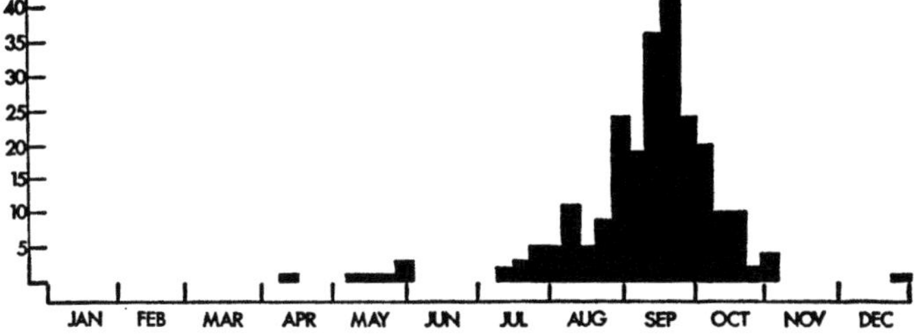

Fig. 65. Seasonal pattern of Pectoral Sandpipers *Calidris melanotos* in Britain and Ireland during 1958–67

and Norfolk each averaging more than two a year and Co. Derry, Co. Cork, the Isles of Scilly, Cornwall and Kent each averaging at least one. Although Norfolk is very well-watched compared with some parts of western Britain and Ireland (see Fig. 1), it is nevertheless noteworthy that more were recorded there than in any other county during the ten years.

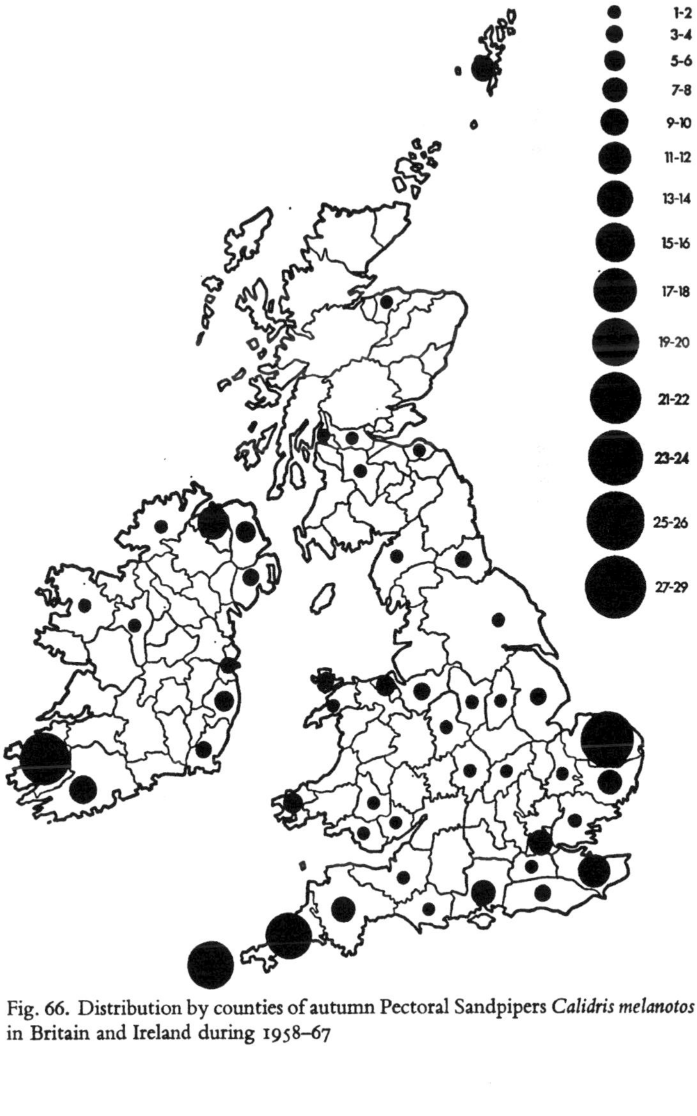

Fig. 66. Distribution by counties of autumn Pectoral Sandpipers *Calidris melanotos* in Britain and Ireland during 1958–67

As a comparison with the Pectoral Sandpipers, we may consider the records of other Nearctic waders during the same period. The species concerned, with the number of records of each in the ten years given in brackets, were Killdeer *Charadrius vociferus* (seven), Lesser Golden Plover *Pluvialis dominica* (seven), Short-billed Dowitcher *Limnodromus griseus* (nine), Long-billed Dowitcher *L. scolopaceus* (ten) and indeterminate dowitchers (30), Stilt Sandpiper *Micropalama himantopus* (five), Upland Sandpiper *Bartramia longicauda* (five), Solitary Sandpiper *Tringa solitaria* (four), Spotted Sandpiper *T. macularia* (five), Greater Yellowlegs *T. melanoleuca* (four), Lesser Yellowlegs *T. flavipes* (29), Least Sandpiper *Calidris minutilla* (nine), Baird's Sandpiper *C. bairdii* (22), White-rumped Sandpiper *C. fuscicollis* (47), Semipalmated Sandpiper *C. pusilla* (seven), Western Sandpiper *C. mauri* (two), Buff-breasted Sandpiper *Tryngites subruficollis* (40) and Wilson's Phalarope *Phalaropus tricolor* (27). There were thus 269 records of these 17 species during 1958–67, a total which is conveniently similar to that of the 243 Pectoral Sandpipers recorded in the same period. The temporal distribution of these records (Fig. 67) may be compared with that of the Pectorals (Fig. 65). Since the records of 17 species are amalgamated, one would not expect to find a clear-cut distribution, and the double peak in autumn is not surprising.

While only seven of the 243 Pectoral Sandpipers occurred in spring, 28 of the 269 other Nearctic waders were observed at that season; but, with a heterogeneous batch of records, comparisons must be drawn with caution. There are insufficient data to justify separate treatment for individual species, though it is clear that the pattern for each is usually distinct (the Killdeers, for instance, all being first recorded in November, February or March). It is, nevertheless, interesting to note that, whereas all seven spring Pectoral Sandpipers were in the west (Ireland, south-west England and Wales), seven of

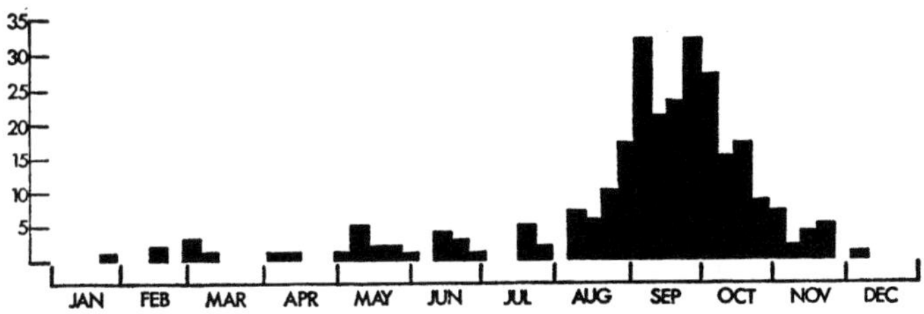

Fig. 67. Seasonal pattern of Nearctic waders (other than Pectoral Sandpipers *Calidris melanotos*) in Britain and Ireland during 1958–67

Fig. 68. Distribution by counties of spring Pectoral Sandpipers *Calidris melanotos* (open circles) and the 17 other Nearctic waders (filled circles) in Britain and Ireland during 1958–67

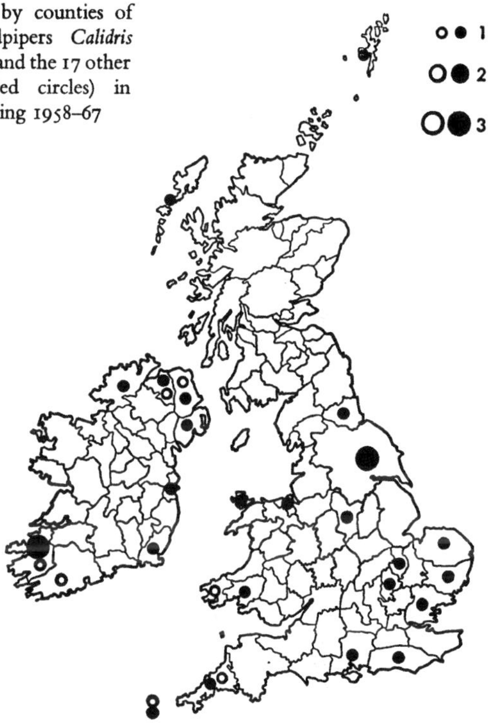

the other Nearctic waders in spring were in the east (north-east and eastern England and East Anglia) and 14 in the west. The difference between these two spring patterns is illustrated in Fig. 68. The fact that 33% of the spring Nearctic waders other than Pectoral Sandpipers occurred in the east suggests that a proportion had crossed the Atlantic during a preceding season and were pursuing a south-north migration, perhaps between Africa and northern Europe. This idea is also supported by the recurrence of what seem likely to be the same individuals at the same localities in successive seasons, for example the single Wilson's Phalaropes seen at Scaling Dam Reservoir, Yorkshire, on 20th and 21st June 1965 and found dead there on 22nd June 1966 (*Brit. Birds*, 59: 290; 60: 320). On the other hand, one should not neglect the fact that all the Pectoral Sandpipers and 66% of the other Nearctic waders in spring were recorded in the west. This suggests that the majority of spring occurrences were due to spring transatlantic crossings.

The distribution of the autumn records of the 17 other Nearctic waders (Fig. 69) is similar to that of the Pectoral Sandpipers (Fig. 66), with averages of more than two a year in Co. Kerry and the Isles of Scilly and more than one a year in Co. Cork and Norfolk. Ignoring records other than those distinctly in the west (Ireland and south-west

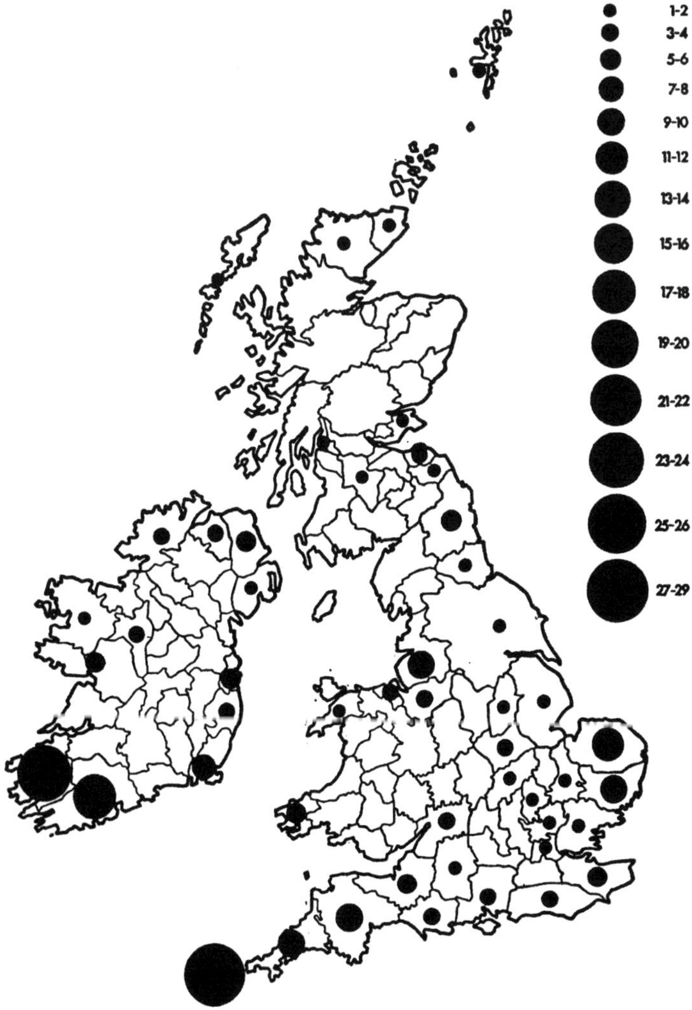

Fig. 69. Distribution by counties of autumn Nearctic waders (other than Pectoral Sandpipers *Calidris melanotos*) in Britain and Ireland during 1958–67

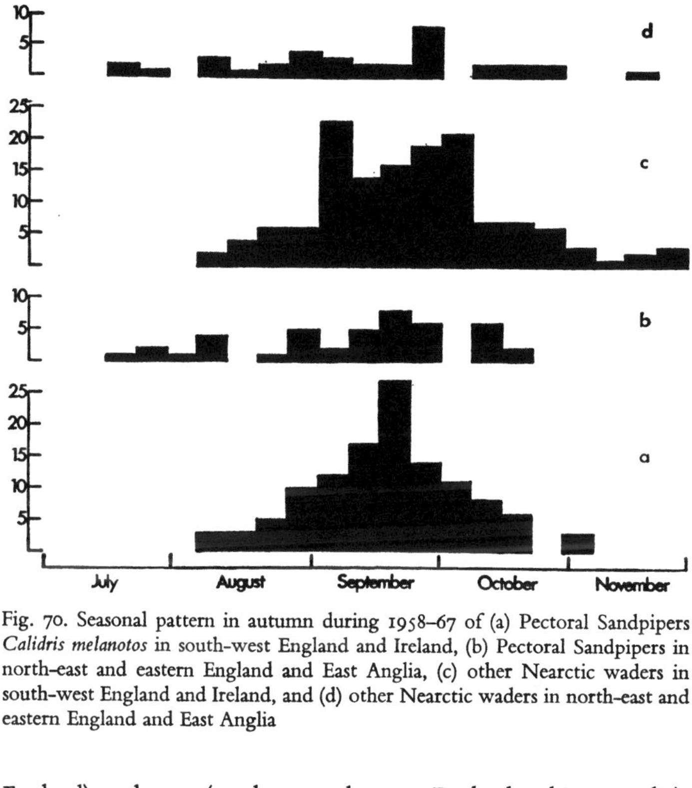

Fig. 70. Seasonal pattern in autumn during 1958–67 of (a) Pectoral Sandpipers *Calidris melanotos* in south-west England and Ireland, (b) Pectoral Sandpipers in north-east and eastern England and East Anglia, (c) other Nearctic waders in south-west England and Ireland, and (d) other Nearctic waders in north-east and eastern England and East Anglia

England) or the east (north-east and eastern England and East Anglia), two contrasting patterns emerge for these two areas when the data are analysed separately (Fig. 70). In the west, Pectoral Sandpiper records form a normal distribution without the negative skewness exhibited by the entire data; this skewness is largely accounted for when the eastern data are examined, as Pectoral Sandpipers occurred earlier in the east than in the west. This is clearly demonstrated by the histograms and also by the fact that 33% of the Pectoral Sandpipers in the east occurred before 3rd September compared with only 17% of those in the west. This pattern is strikingly unexpected for a species assumed to arrive here in autumn as a transatlantic vagrant; one might have anticipated a coincident or subsequent peak in the east, reflecting simultaneous arrival or onward passage from the west.

The records of the 17 other Nearctic waders, similarly split into eastern and western components (Fig. 70), help to shed light on this problem. Again bearing in mind the heterogeneous nature of the data, the patterns are strikingly similar to those of the Pectoral Sandpipers. The data from the east of Britain do not exhibit a normal distribution and, like the Pectoral Sandpipers, these other Nearctic waders appear there earlier, 37% of those in the east being recorded before 3rd September compared with only 13% of those in the west. The small numbers would make separate analysis of each species invalid, but the White-rumped Sandpiper may be taken as an example since this was the second commonest Nearctic wader in Britain and Ireland in 1958–67. The total of 47 individuals included 45 in autumn. Five of the nine in the east of Britain occurred before 3rd September, compared with only two of the 25 in the west. This is the same pattern as exhibited by the Pectoral Sandpipers and the 17 other Nearctic waders treated as a whole. Rogers (1972) has suggested that these birds are likely to be failed breeders, newly arrived as true vagrants from America, and that they are probably adults. They are, indeed, likely to be mostly adults (though the data are insufficient to confirm this), but the east coast bias suggests that they are not new arrivals. The logical conclusion must be that the majority of these early east coast birds in the coastal counties from Northumberland to Essex (and probably those in the east of Scotland, the Midlands and south-east England, though these were not included in the comparisons) had crossed the Atlantic during a previous season and were undertaking a north-south migration from northern Europe.

Even by a simple visual comparison of Figs. 66 and 69, however, it is evident that Pectoral Sandpipers are relatively more common in the east of Britain than are other Nearctic waders. In fact, more than a quarter of the autumn Pectoral Sandpipers were seen in the British east coast counties compared with less than a fifth of the other Nearctic waders. This disproportionately high percentage of Pectoral Sandpipers leads one to speculate whether *some* of those on the east coast may have been of Siberian origin, for this species breeds not only in North America west to Alaska, but also on the arctic coast of Siberia from the eastern side of the Taimyr Peninsula east to the delta of the Kolyma (Vaurie 1965). Of the 17 other Nearctic waders used as a control, however, no less than four more—Lesser Golden Plover, Long-billed Dowitcher, Western Sandpiper and Baird's Sandpiper—also have breeding distributions extending into Siberia. These four combined show a pattern similar to that of the Pectoral Sandpiper, with 25% of the records in the east, but individual analyses established that the first three all conform to the pattern of the remaining 13 strictly Nearctic species as a block. The notable exception, sufficiently marked to raise the combined average of these four to such an extent, was Baird's Sandpiper which had ten in the west and no less than seven (41%) in the east. This is, of course, based on a very small sample (only 22 individuals in the whole of Britain and Ireland), but the addition of three 1968 records and five 1970 records hardly changes the proportion (36% in the east). Standard χ^2 tests were applied to these figures to examine the probability (p) that the differences were accidental and it was found that $p = 0{\cdot}050$ for Pectoral Sandpiper and $p = 0{\cdot}025$ for Baird's Sandpiper (compared in both cases with the remaining 16 Nearctic waders). Similar tests were applied to each of these other 16 (comparing them individually with the remaining 15) and in each case $p > 0{\cdot}100$, even for the White-rumped Sandpiper which had the high east coast percentage of 26%.

While these differences between the west-east proportions of Baird's and Pectoral Sandpipers on the one hand and all other Nearctic waders combined on the other are only slight, the data do suggest a genuine differentiation. This could be due to a variety of reasons: these two species could be better adapted in some way to survival in Europe or Africa so that cumulatively there was a higher proportion regularly migrating north and south through Britain (though the spring records suggest that, if anything, the reverse is the case); they might be more likely for some reason to overfly Ireland and western Britain and occur in higher numbers in the east; there might be some observer bias resulting in their being selectively overlooked or the other species being particularly noticed in western districts (or vice versa in the east); habitat differences between the west and the east might result in segregation; and other explanations could no doubt be suggested. These all seem unlikely, however, and a more logical explanation appears to be that a proportion of the Pectoral and Baird's Sandpipers arriving in Britain come from the east, not the west, and are derived from the Siberian populations. It is highly unlikely that this hypothesis will ever be proved, but it is a theory which accords with the patterns shown by the 1958–67 data. The figures available suggest that perhaps as many as one in six of the Baird's Sandpipers and one in 14 of the Pectoral Sandpipers occurring in Britain and Ireland in autumn may be of eastern rather than western origin; but this is highly speculative.

Turning now to the annual totals (Fig. 71), Pectoral Sandpipers showed very distinct autumn peaks in 1961 and 1967. Although marginally more (42) were seen in autumn 1967, the 41 records in autumn 1961 probably represented a higher number reaching Britain

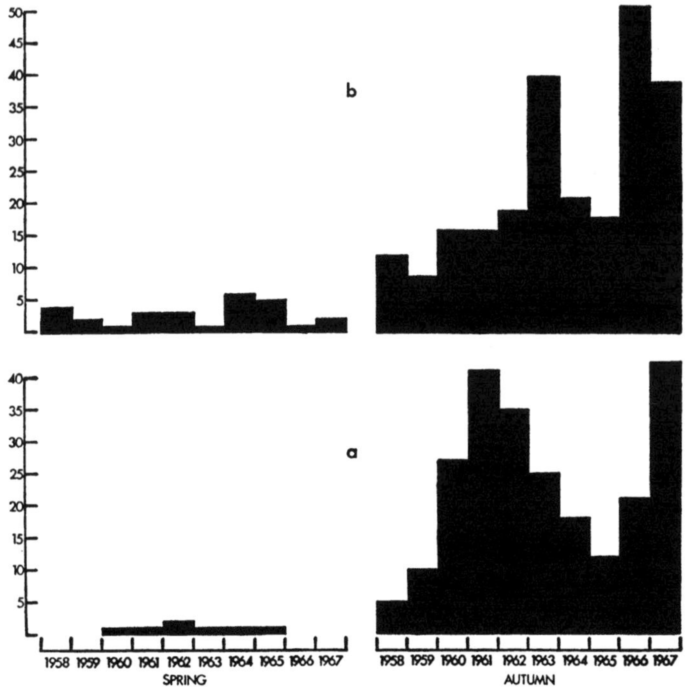

Fig. 71. Annual pattern of (a) Pectoral Sandpipers *Calidris melanotos* and (b) other Nearctic waders in Britain and Ireland during 1958–67 with the spring and autumn records shown separately

and Ireland, allowing for the increase in observers by the later year. The records over the ten years have the appearance of a cyclical pattern and, indeed, the dominance of Pectoral Sandpipers in the British and Irish records of American waders in the early 1960s led Williamson (1963) to suggest that the species was enjoying considerable success at that time.

The pattern of records on this side of the Atlantic may well reflect population levels in North America, especially since the Pectoral Sandpipers are not in phase with the other 17 Nearctic waders as a whole, which had peaks in the autumns of 1963 and 1966. It might reasonably be expected that the number of Nearctic waders appearing here in any particular autumn was determined to a considerable extent by the North Atlantic weather situation (e.g. Williamson and Ferguson-Lees 1960) and that a 'good' year for Nearctic waders as a whole would also be a 'good' year for Pectoral Sandpipers. Clearly, however, the timing of the migration of each species within North America will be distinct and, therefore, the timing and latitude of suitable meteorological conditions for transatlantic crossings will be critical, leading to selective vagrancy to Europe each year. It is also doubtless misleading to treat 17 species in bulk in this sort of comparison, for the individual population levels are unlikely to be in phase. The numbers of most of the 17 other Nearctic waders are too small to warrant individual treatment, but the three commonest (White-rumped Sandpiper, Buff-breasted Sandpiper and Lesser Yellowlegs) may be taken as examples (Fig. 72). The differences in the patterns of the Pectoral Sandpiper and each of these three species are no doubt due partly to the selective action of the timing of suitable meteorological conditions for transatlantic vagrancy as well as to the differing statuses of the populations.

The annual patterns are only partially reflected in the east of Britain (Fig. 73). In years when large numbers of Pectoral Sandpipers cross the Atlantic and reach western areas, there is obviously good penetration to the east (as in 1961 and 1967). Above average proportions of the Pectoral Sandpipers occurred in the east, however, in 1959, 1960, 1962 and 1965. The 1962 numbers seem very likely to have been an aftermath of the previous autumn's invasion, consisting partly of birds which had spent twelve months on this side of the Atlantic. It is conceivably not irrelevant, also, that 1962 was the only year in which more than one Pectoral Sandpiper was recorded in spring (though it has already been noted that all seven spring individuals were in the west and therefore suggestive of spring transatlantic crossings). In the other three autumns, however, the disproportionate numbers of Pectoral Sandpipers in the east did not follow large numbers in the west in the previous year, but they coincided with above average numbers in Britain and Ireland of three eastern waders—Little Stint

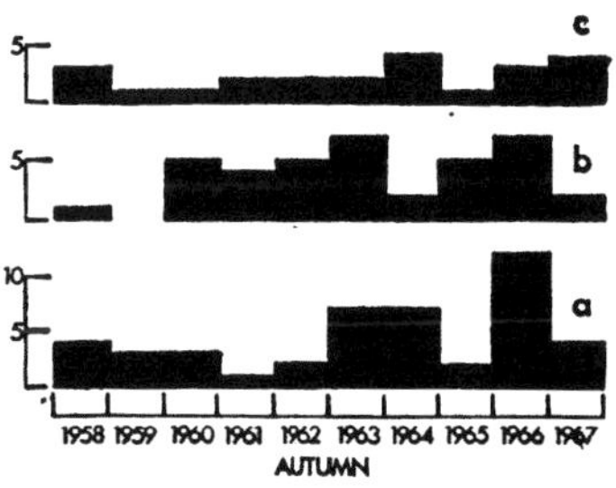

Fig. 72. Annual patterns of autumn records of (a) White-rumped Sandpipers *Calidris fuscicollis*, (b) Buff-breasted Sandpipers *Tryngites subruficollis*, and (c) Lesser Yellowlegs *Tringa flavipes* in Britain and Ireland during 1958–67

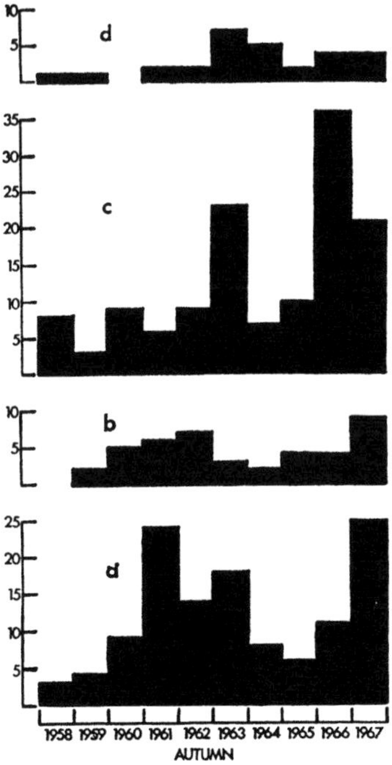

Fig. 73. Annual patterns of autumn records during 1958–67 of (a) Pectoral Sandpipers *Calidris melanotos* in south-west England and Ireland, (b) Pectoral Sandpipers in north-east and eastern England and East Anglia, (c) other Nearctic waders (excluding Baird's Sandpipers *C. bairdii*) in south-west England and Ireland, and (d) other Nearctic waders (again excluding Baird's Sandpipers) in north-east and eastern England and East Anglia

Calidris minuta, Temminck's Stint *C. temminckii* and Curlew Sandpiper *C. ferruginea* (see Table 4). It is tempting to suggest that these may have been years when some Pectoral Sandpipers arrived in Britain from the east.

The east-west pattern of the other Nearctic waders (excluding Baird's Sandpiper) is somewhat confirmatory (Fig. 73), for the only year in which the proportion in the east was significantly above normal was 1964, following the high numbers in the previous autumn. It is perhaps also significant that the highest spring numbers in the ten years were in 1964 and 1965 and, even though the majority of these may have been recent transatlantic arrivals, there were four in the east compared with only three in the west. The species involved in the subsequent seasons were those that had occurred in numbers during the peak autumn. It is, however, surprising that there were relatively so few in 1967, following the 1966 influx; both the spring numbers and the proportion of autumn records in the east were below average in that year.

To summarise, the 1958–67 records of Pectoral Sandpiper suggest that (1) those which cross the Atlantic in autumn make landfall mainly in Ireland and south-west England, though small numbers penetrate through to eastern England; (2) spring occurrences appear likely to be mainly new transatlantic arrivals; (3) after years with a heavy autumn passage it is probable that some individuals recur in the following seasons in the course of north-south migration on this side of the Atlantic; (4) there is evidence of the possibility of vagrancy of very small numbers from the east. The first, third and fourth of these conclusions apply also to Baird's Sandpiper, and the first three conclusions apply to the other Nearctic waders which occur in Britain and Ireland.

Sabine's Gull
Larus sabini

Thirteen Sabine's Gulls in September 1950 were considered at the time to constitute an invasion (Anon 1951). Yet a total of 203 was recorded in Britain and Ireland during 1958–67 (including twelve seen from fishing boats or ships in inshore waters), there being an average of 15 a year over nine of the years and one exceptional influx in the tenth. *The Handbook* drew special attention to the great rarity of adults in the records up to 1940, so it is of interest that the published details in 1958–67 include 55 adults and 81 immatures. One may not be justified in deducing that 40% of the Sabine's Gulls now occurring are adults, however, for the comment in *The Handbook* may have led some county report editors to mention adults specifically (*cf.* red-spotted and white-spotted Bluethroats in Chapter 4) and the 67 individuals for which no age data were given may, in consequence, have included a higher proportion of immatures.

Only seven Sabine's Gulls were reported in spring (March–June), but, while the others were spread from July to December, 72% of all the records were in eight weeks during September and October (Fig. 74). There was an exceptional influx in the autumn of 1967 (Fig. 75), with the main arrival during the first few days of September, but, even if the records for that year are omitted, 65% still fell in September–October and the median shifts only from 23rd to 27th September.

It is a great pity that the ages of a third of the individuals are not available for analysis, since the data which were published show interesting patterns (Fig. 76). Taking all the records, 31% of the adults occurred before 3rd September compared with only 4% of the immatures (just three individuals). If the exceptional 1967 records are excluded, the pattern is even more striking, with as many as 41% of the adults before 10th September compared with only 11% of the immatures. This pattern is also found in the Bay of Biscay, where most of the Sabine's Gulls in August are adults, and Mayaud (1961) has suggested that these are failed breeders. As well as appearing earlier than the immatures in Britain and Ireland, the adults also tended to occur later for, again excluding the 1967 influx, only 14% of the

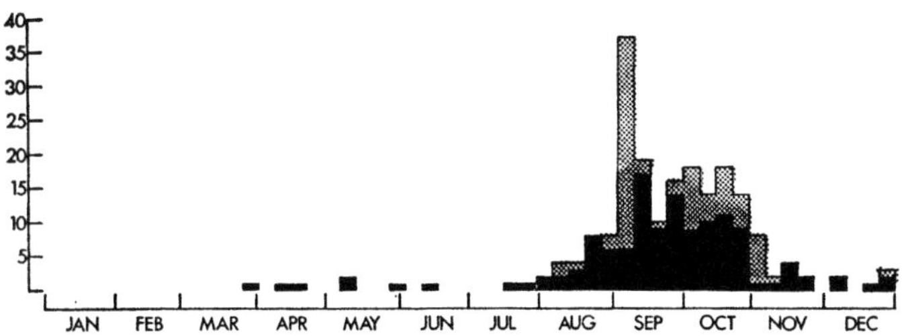

Fig. 74. Seasonal pattern of Sabine's Gulls *Larus sabini* in Britain and Ireland during 1958–67; the 1967 records are shown shaded

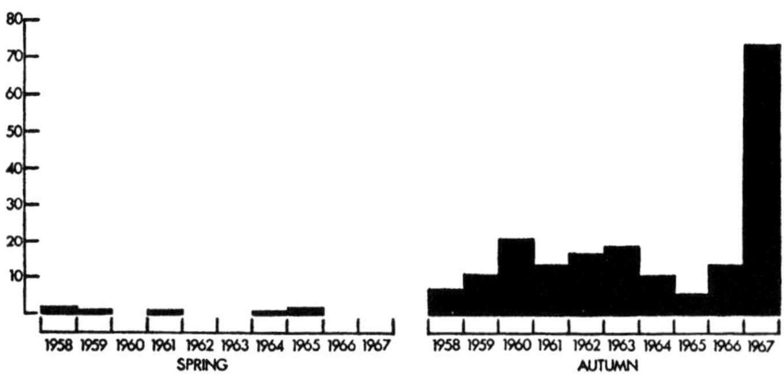

Fig. 75. Annual pattern of Sabine's Gulls *Larus sabini* in Britain and Ireland during 1958–67 with the spring and autumn records shown separately

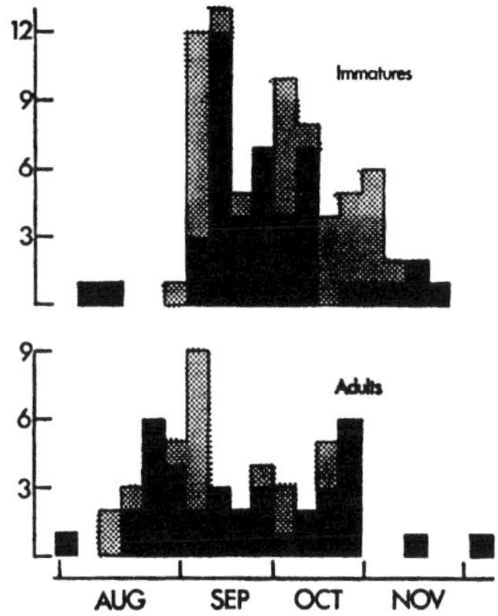

Fig. 76. Seasonal patterns of adult and immature Sabine's Gulls *Larus sabini* in Britain and Ireland in autumn during 1958–67; the 1967 records are shown shaded.

immatures were recorded after mid-October, compared with 30% of the adults.

The seven spring records were confined to four counties—three in Kent, two in Dorset and one each in Co. Antrim and Co. Wexford. *The Handbook* noted that most autumn records up to 1940 were in Yorkshire and Norfolk; it seems likely that this pattern was merely a reflection of the distribution of observers at that time for, of the autumn records in 1958–67, almost 40% were in Cornwall—the vast majority at one locality (St Ives Island)—and only three other counties, all in the west (Co. Cork, Co. Kerry and Dorset), averaged more than one per year (Fig. 77).

The experience of seawatchers in these western areas is that most Sabine's Gulls occur during or immediately after westerly or north-westerly gales (e.g. Phillips 1966) and they are clearly the result of displacement eastwards, especially into the 'funnel' of the Bristol Channel. The early adults (17 individuals before 3rd September) conformed to the general pattern, with most in the west, but there were seven in Co. Cork compared with five in Cornwall. Unlike the Nearctic waders (Fig. 70), this species was not recorded at significantly different times on the west coast (Ireland and south-west England) and the east coast (Northumberland to Essex).

The exceptional influx in autumn 1967 has already been mentioned but it should be noted that it involved at least 74 individuals. In that year immatures occurred both earlier and later than usual and the influx, as well as being exceptionally large, was atypical in that the proportion of adults accompanying the early immatures was unusually high (Fig. 76). The numbers of this species in each of the other autumns (Fig. 75) varied from six (1958) to 21 (1960). Whereas the patterns of Pectoral Sandpipers could be compared with those of other Nearctic waders, there are too few records of other Nearctic gulls to provide a

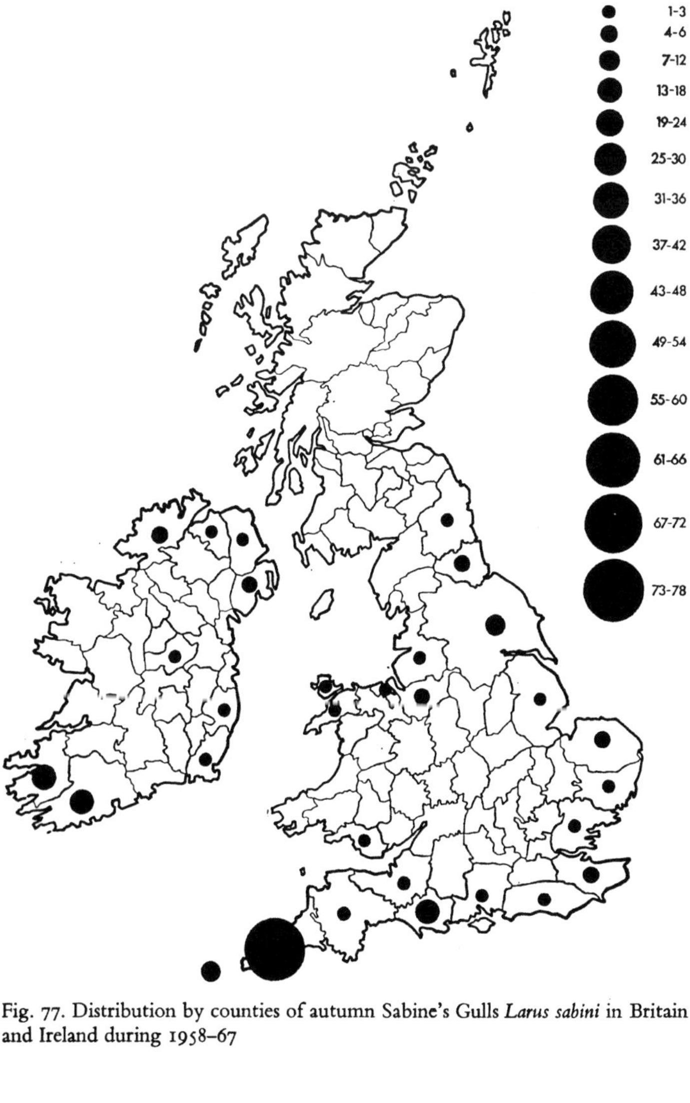

Fig. 77. Distribution by counties of autumn Sabine's Gulls *Larus sabini* in Britain and Ireland during 1958–67

good comparison with those of Sabine's Gulls. The only records were six Bonaparte's Gulls *Larus philadelphia*, two Laughing Gulls *L. atricilla* and one Kumlien's Gull *L. glaucoides kumlieni*. Apart from a Bonaparte's in June, a Laughing in May and the Kumlien's in January, the other six were all in autumn (August to November). There were singles in 1961 and 1963, and the remaining four were all in 1967, thus coinciding with the peak year for Sabine's Gulls. It will be recalled (Fig. 71) that this was also the peak year for Nearctic waders, including Pectoral Sandpipers. The third highest total of Sabine's Gulls, 19 in 1963, coincided with another peak of Nearctic waders, but the latter were below average in 1960, the year of the second highest total.

Sabine's Gull is nowhere a common breeding species but it has a Holarctic distribution extending from north-west Greenland westwards through Baffin Island, Arctic Canada, Alaska and Siberia to the Taimyr Peninsula, with irregular nesting further west to Spitsbergen. As recently as the 1950's, the wintering area was considered to be in the Gulf of Gascony (Fisher and Lockley 1953, Ferguson-Lees 1955). In the Bay of Biscay the numbers decline through the autumn—a recent example being 1,000 in the area of Belle Ile to Le Croisic in late August 1965, 100 in late September and 40 in early October (Ricard 1966)—and, though the disappearance from this region by mid-December was thought to be puzzling, it was suggested that it might represent an early return northwards to the breeding areas. Observations of passage in a narrow line near the coasts of Senegal and Morocco during May, the birds moving northwards very rapidly and hardly stopping to feed (Roux 1961), provided one of the first clues to the winter quarters which have now been shown to lie off south-west and south Africa between the Bay of Lüderitz and Port Elizabeth (Morgan and Wheeler 1958, Liversidge and Courtenay-Latimer 1963, Zoutendyk 1965, Mayaud 1965). The birds in their winter quarters and the spring passage northwards along the west African coast are now regularly observed (e.g. Bourne 1970).

The east Siberian and Alaskan populations pass through the Bering Strait and down the east Pacific coast and, because no significant numbers are observed along the coasts of Greenland or Norway, both Mayaud (1961) and Bourne (1965) deduced that the west Siberian population joined those from the east and also passed into the Pacific. The relative scarcity of British east coast records compared with those of, say, Long-tailed Skuas (Fig. 49), as well as the total absence of Scottish records in the ten years, makes it unlikely that the British and Irish records involve west Siberian birds passing between Iceland

and Shetland through the Norwegian Sea. The pattern makes it far more probable that the Sabine's Gulls observed here are Nearctic ones which have passed along the North American east coast south to Maine and Massachusetts, crossed the Atlantic south of Greenland towards Brittany and then been displaced by westerly winds to Ireland and western Britain.

American land-birds

All the records of American land-birds in western Europe up to 1953 (including dubious and officially rejected records as well as those accepted in national check-lists) were brought together and discussed by Alexander and Fitter (1955) who thus provided an indispensable base-line. The 29 American passerines recorded in Britain, Ireland and France during 1951–62 were thoroughly analysed by Nisbet (1963) who also compared the patterns with earlier records. These two papers are essential reading for students of this subject and the brief treatment here in no way attempts to replace them.

During 1958–67 there were 52 records of American passerines (including two identified only to genus) and five of American cuckoos in Britain and Ireland. A further 15 passerines were recorded in 1968 and so, in order to provide a larger sample, this section is concerned with the eleven years 1958–68 (unlike all others in this book, which are strictly confined to the ten years 1958–67). The 1968 records have all been published in one place (Smith 1969) and so the reader may easily determine the 1958–67 patterns from the 1958–68 ones.

Table 7 shows the American land-birds recorded in 1958–68 and also during two preceding periods. Leaving aside the three non-passerines at the top of the list, it demonstrates the very striking increase in observations of Nearctic passerines in the most recent eleven-year period—nearly four times as many as the total for all time up to then and more than $5\frac{1}{2}$ times as many as in the previous eleven years. It is impossible, however, to know the extent to which this increase reflects the annually growing number of observers and also their greater appreciation of the possibility of finding American land-birds (indeed, the preoccupation of some of them with doing so). Nisbet (1963) drew attention to the fact that the earlier records might be biased towards the larger and more conspicuous species, and the proportion of the three non-passerines has fallen from 75% up to 1946 to 45% in 1947–57 and 7% in 1958–68. The drop in the actual numbers (especially of the Yellow-billed Cuckoos*), however, from 18 in the first period to ten

Table 7. Summary of all accepted records of American land-birds in Britain and Ireland

In addition to the species listed below, there was an unidentified wood warbler *Dendroica sp*, an unidentified tanager *Piranga sp* and an Indigo Bunting *Passerina cyanea* (a common cage-bird) in 1958–68, and the wing of an Ovenbird *Seiurus aurocapillus* was found on the tideline on 4th January 1969. The first two of these, but not the last two, have been included in all the totals in this section. Four records of the American Pipit *Anthus spinoletta rubescens* have been excluded since the breeding range extends into Greenland

	Up to 1946	1947–57	1958–68
Yellow-billed Cuckoo *Coccyzus americanus*	15	6	3
Black-billed Cuckoo *C. erythrophthalmus*	2	2	2
Nighthawk *Chordeiles minor*	1	2	0
Horned Lark *Eremophila alpestris alpestris*	0	1	0
Brown Thrasher *Toxostoma rufum*	0	0	1
American Robin *Turdus migratorius*	3	4	9
Olive-backed Thrush *Hylocichla ustulata*	0	1	2
Grey-cheeked Thrush *H. minima*	0	1	6
Red-eyed Vireo *Vireo olivaceus*	0	1	6
Black-and-white Warbler *Mniotilta varia*	1	0	0
Parula Warbler *Parula americana*	0	0	3
Yellow Warbler *Dendroica petechia*	0	0	1
Myrtle Warbler *D. coronata*	0	1	2
Blackpoll Warbler *D. striata*	0	0	2
Northern Waterthrush *Seiurus noveboracensis*	0	0	2
Yellowthroat *Geothlypis trichas*	0	1	0
American Redstart *Setophaga ruticilla*	0	0	2
Bobolink *Dolichonyx oryzivorus*	0	0	2
Baltimore Oriole *Icterus galbula*	0	0	10
Summer Tanager *Piranga rubra*	0	1	0
Song Sparrow *Melospiza melodia*	0	0	2
Fox Sparrow *Passerella iliaca*	0	0	1
White-throated Sparrow *Zonotrichia albicollis*	1	0	7
Slate-coloured Junco *Junco hyemalis*	1	0	3
Rufous-sided Towhee *Pipilo erythrophthalmus*	0	0	1
Rose-breasted Grosbeak *Pheucticus ludovicianus*	0	1	3
TOTALS	24	22	70

* Scientific names of species in this section are given in Table 7.

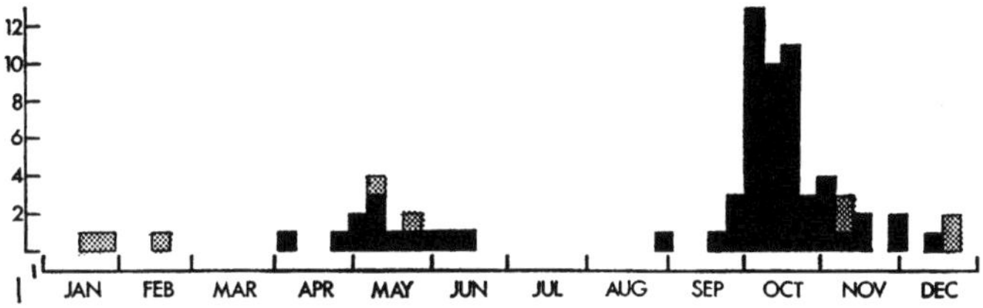

Fig. 78. Seasonal pattern of American land-birds in Britain and Ireland during 1958–68; the records of American Robins *Turdus migratorius* are shown shaded

in the next eleven years and then to only five in the last eleven, clearly represents a genuine decrease. Hereafter, the cuckoos are included with the passerines, so totals relate to all land-birds.

Taking the spring records and autumn (and winter) records separately (April–June and August–February*: see Fig. 78), the annual totals varied from none to four in spring and none to 14 in autumn:

	1958	*1959*	*1960*	*1961*	*1962*	*1963*	*1964*	*1965*	*1966*	*1967*	*1968*
Spring	0	1	1	3	0	0	1	1	4	2	1
Autumn	3	0	2	2	6	4	3	6	5	13	14

The majority of the autumn records were concentrated in a few well-watched localities in the south-west of Britain and Ireland—the Isles of Scilly, Lundy (Devon), Skokholm (Pembrokeshire), Bardsey (Caernarvonshire) and Cape Clear Island (Co. Cork)—which accounted for over three-fifths of the autumn records in 1958–68, nearly half of them in Scilly (Fig. 79b). Bonham (1970) has mapped the 1951–68 records (spring and autumn combined) by individual localities.

The distribution of the records through the year (Fig. 78) shows very distinct peaks, in April to mid-June (especially early May) and in late September to mid-November. These are even more marked if the nine American Robins—two in May and the others in November–February (thus comparable with the Killdeers)—are excluded. A fifth of the records fell in the spring period and three-quarters in the autumn, with no less than 71% of the latter in the first three weeks of October. This concentration is even more marked when single years are examined. Three American passerines have occurred within one week once, four within one week on two occasions, and five and seven

*January/February records of American Robins in 1965 (Kerry) and 1966 (Dorset and Surrey) have all been treated as referring to the previous autumn.

within one week once each (Table 8). (It is a strange coincidence, but probably nothing more, that when American land-birds reached Skokholm they were always the first of the influx and, conversely, those on Cape Clear Island were always the last.) Several of the species also show a remarkable concentration within a short period. Six of the seven Red-eyed Vireos occurred within just three days (4th–6th October) in five different years; three of the four Rose-breasted Grosbeaks also appeared within three days (5th–7th October) in three

Table 8. The five most concentrated arrivals of American land-birds in Britain and Ireland

Two records which fell in the following or preceding week are included in brackets for completeness. Scientific names are given in table 7

Date	Species	Locality	TOTALS
1962			
4th October	Red-eyed Vireo	St Agnes, Isles of Scilly	
4th October	Red-eyed Vireo	St Agnes	
5th October	Baltimore Oriole	Beachy Head, Sussex	4
7th October	Rose-breasted Grosbeak	Cape Clear Island, Co. Cork	
1967			
5th October	Rose-breasted Grosbeak	Skokholm, Pembrokeshire	
5th October	Baltimore Oriole	Skokholm	3
6th October	Red-eyed Vireo	Cape Clear Island	
14th October	Olive-backed Thrush	Skokholm	
14th October	Red-eyed Vireo	Skokholm	
15th October	White-throated Sparrow	Bardsey, Caernarvonshire	
17th October	Baltimore Oriole	Lundy, Devon	7
17th October	Baltimore Oriole	Lundy	
18th October	Baltimore Oriole	St Agnes	
19th October	Black-billed Cuckoo	Lundy	
(21st October	American Redstart	Porthgwarra, Cornwall)	
1968			
(6th October	Red-eyed Vireo	St Agnes)	
9th October	Parula Warbler	Portland Bill, Dorset	
10th October	Bobolink	St Mary's, Isles of Scilly,	
12th October	Blackpoll Warbler	St Agnes	5
13th October	American Redstart	Cape Clear Island	
14th October	Olive-backed Thrush	Cape Clear Island	
17th October	Grey-cheeked Thrush	Horden, Co. Durham	
19th October	White-throated Sparrow	Beachy Head	
22nd October	Blackpoll Warbler	Bardsey	4
23rd October	Myrtle Warbler	St Mary's	

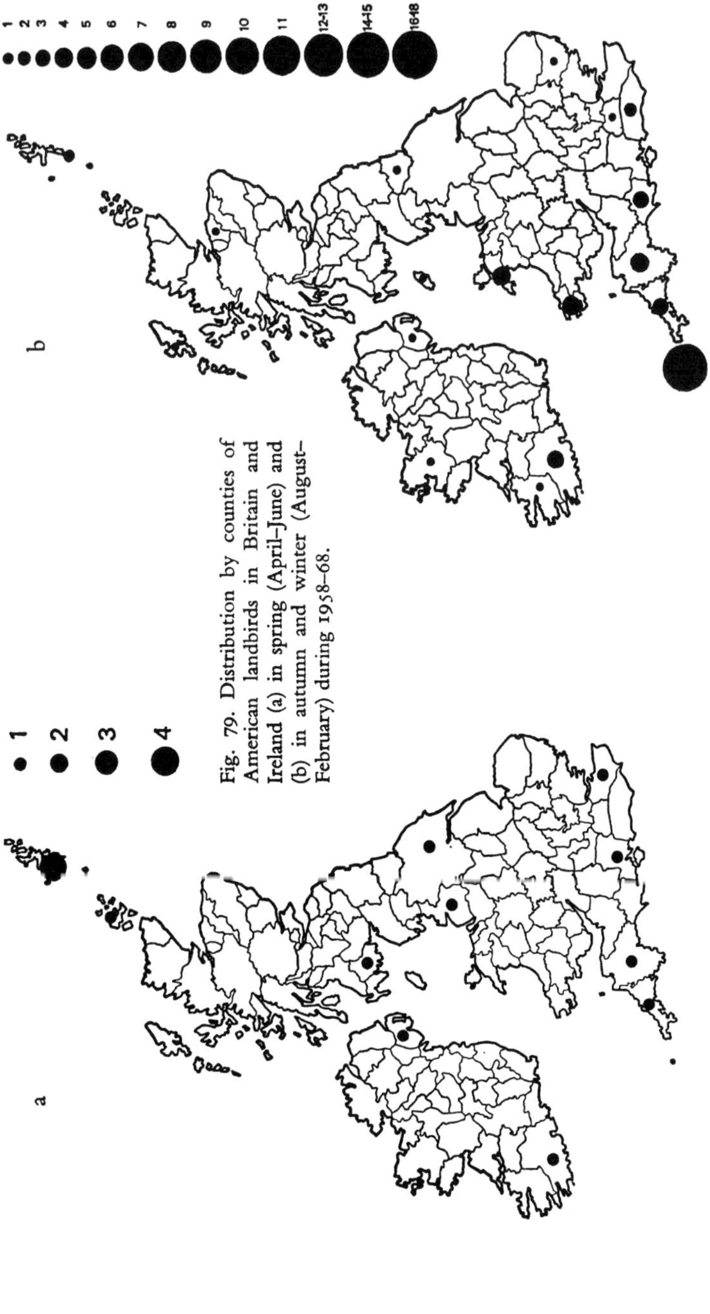

Fig. 79. Distribution by counties of American landbirds in Britain and Ireland (a) in spring (April–June) and (b) in autumn and winter (August–February) during 1958–68.

Table 9. Seasonal distribution of the records of each family of American land-birds recorded in Britain and Ireland in 1958–68

	SPRING Apr–Jun	AUTUMN Aug–Nov	WINTER Dec–Feb
Cuculidae (cuckoos)	0	4	1
Mimidae (thrashers and mockingbirds)	0	1	0
Turdidae (thrushes, except next)	0	8	0
American Robin *Turdus migratorius*	2	2	5
Vireonidae (vireos)	0	6	0
Parulidae (wood warblers)	0	13	0
Icteridae (blackbirds and orioles)	1	11	0
Thraupidae (tanagers)	0	1	0
Fringillidae (finches)	1	3	0
Emberizidae (buntings and sparrows)	10	3	0
TOTALS	14	52	6

different years; and the two autumn Olive-backed Thrushes both occurred on 14th October in different years. Since the peak of autumn arrivals extends over three weeks, these patterns are clearly not due to coincidence and, while further data are required for most species, it seems probable that the British and Irish records are reflecting the North American migration patterns of the individual species. No light is thrown on the vexed question of assisted passage—whether most of these passerines cross the Atlantic on board ships—by the concentration of birds in a short period of time in a small area of western Britain and Ireland and coincidentally with westerly winds. These can all be explained equally by unassisted arrival on the westerly winds or by assisted arrival on transatlantic ships following easterly displacement by the westerly winds off the North American coast. The classic example of the latter was described by Durand (1963), who has also (1972) surveyed other records on transatlantic ship-crossings.

When the spring records are considered, it is very striking that 71% refer to buntings or 'sparrows' (Emberizidae), compared with only 6% of those in autumn (Table 9). No wood warblers (Parulidae) or vireos (Vireonidae) were identified in spring, although together they made up 37% of the autumn records. It is not merely a case of the mainly seed-eating Emberizidae surviving better on this side of the Atlantic and recurring in Britain and Ireland on spring migration (the mainly insectivorous wood warblers having died), since there were actually more Emberizidae recorded in spring than in autumn, not merely a higher proportion. Nevertheless, this is clearly a part of the explanation, for the spring records (Fig. 79a) were not concentrated in the west like those in autumn (Fig. 79b). It will be recalled that five western localities accounted for 62% of the autumn records; in spring these five localities provided only 14%.

Nisbet (1963) drew attention to the discrepancy between the proportions of American Emberizidae successfully crossing the Atlantic on board ship and those seen in Europe in autumn. He noted that, of the 17 American land-birds recorded as having successfully reached Europe by ship assistance, only three were wood warblers, compared with eleven Emberizidae. The numbers recorded in Britain and Ireland in the autumns of 1958–68 were almost the reverse of this ratio, with 19 wood warblers and vireos and only three Emberizidae. The numbers recorded in spring, however, accorded more closely with the ship pattern.

Tuck (1970) has listed the species most often recorded landing on ships passing along the eastern seaboard of North America. The three most numerous of these in each of his two-month periods were as follows (in descending order of frequency):

APRIL–MAY	Slate-coloured Junco, White-throated Sparrow, Song Sparrow
AUGUST–SEPTEMBER	American Pipit, Myrtle Warbler, Magnolia Warbler *Dendroica magnolia*
OCTOBER–NOVEMBER	Slate-coloured Junco, White-throated Sparrow, White-crowned Sparrow *Zonotrichia leucophrys*

The three species commonest on ships in April–May made up 64% of the spring records of American passerines in Britain and Ireland in 1958–68. On the other hand, the six species commonest on ships in August–November made up less than 12% of the autumn records here, two of them have never been recorded in Britain or Ireland (Magnolia Warbler and White-crowned Sparrow) and one has never been recorded in autumn (Slate-coloured Junco), even though it is the commonest of all on ships in the W. Atlantic in October–November.

The conclusions reached by Nisbet (1963) are confirmed by subsequent events. There is a clear indication that the majority of American land-birds, other than seed-eating Emberizidae, reach western Britain and Ireland naturally. While some may receive ship assistance for the whole or part of their journey, it is most unlikely that this is a major contribution to their vagrancy pattern here. On the other hand, there is a strong likelihood that the occurrences of the seed-eating Emberizidae, particularly in spring, result largely from ship-assisted passage.

CHAPTER 6

Greenish Warbler, Arctic Warbler and Scarlet Rosefinch

This chapter concerns two species which, within Europe, are largely confined to the area south and east of the Baltic (north Germany and Poland, the Baltic States, south Finland and Russia, Figs. 82 and 88), but whose breeding distributions extend almost across the Palearctic; they have in common, too, a tendency to westerly range expansions in the last 30 years or more. For comparison, the rather fewer records of a third species (Arctic Warbler) with a more northerly distribution in Fenno-Scandia, and thence right across the Palearctic, are briefly considered at the same time.

The genus *Phylloscopus*, to which both Greenish and Arctic Warblers belong, is an intriguing group to ornithologists, partly because many of the 30 or so species are very similar and their identification sets observers a challenge. Greenish Warblers are the same size as our common Chiffchaff *Phylloscopus collybita* and Willow Warbler *P. trochilus*, and Arctic Warblers are slightly larger. Although similarly greenish-brown above, they are usually whiter below than our common species and also differ in having more prominent supercilia and small whitish wing-bars (Arctic sometimes shows two on each wing). The best distinguishing marks of these two difficult species are probably leg colour (dark in Greenish, pale in Arctic), head shape (flat and pointed in Arctic) and call-note. On passage in Britain and Ireland they are likely to be found in trees, bushes or low cover near the coast, especially at certain favoured island observatories.

Scarlet Rosefinches are about the same size as House Sparrows *Passer domesticus* and the adult males are, as the vernacular name suggests, largely rose-coloured—on the head, breast and rump—but with browner back and brown wings and tail. Such birds seldom occur here, however, and most seen in Britain and Ireland are lacking pink, being brownish and streaked, with few distinguishing marks except for a staring black eye, two pale wing bars and a dumpier appearance than most other finches and buntings. Those that occur here are often found in stubble fields or feeding on the seed-heads of weeds.

Greenish Warbler
Phylloscopus trochiloides
Arctic Warbler
P. borealis

There was only one record of a Greenish Warbler in Britain and Ireland before 1945 (see *The Handbook*), but then eleven in the next 13 years and no fewer than 46 during 1958–67. The increase in the number of observers, the improvement in their ability to identify difficult warblers and the spread of trapping and ringing must all be borne in mind, but, as Ferguson-Lees (1955) and Williamson (1962) have both pointed out, the rise in the records has coincided with a westward range expansion in Europe (Valikangas 1951a and b, also Palm 1952, Lundberg *et al.* 1954, Merikallio 1958, Mikelsvaar 1963, Veroman 1963, Nørregaard 1964, Borgström 1967, etc.). Unlike Scarlet Rosefinches (see next section), however, there has not been a recent sudden upsurge of records, with an average of only four per year in 1968–72, compared with the five per year in 1958–67.

Greenish Warblers were recorded in every month from June to January during the ten years, but most were in September or October and more than three-quarters during the ten weeks from 27th August to 4th November (Fig. 80). Few were recorded in spring and summer, there having been only three in June (two in Yorkshire and one in Kent) and one in late July (Isle of Man) before the autumn arrivals in August onwards. Surprisingly, in view of the fact that the normal wintering area is India, there were two winter records in the ten years—1st January to 26th February 1961 in Middlesex and 20th December

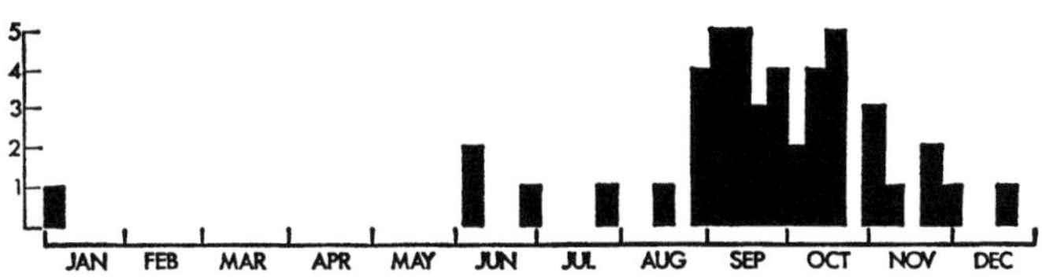

Fig. 80. Seasonal pattern of Greenish Warblers *Phylloscopus trochiloides* in Britain and Ireland during 1958–67

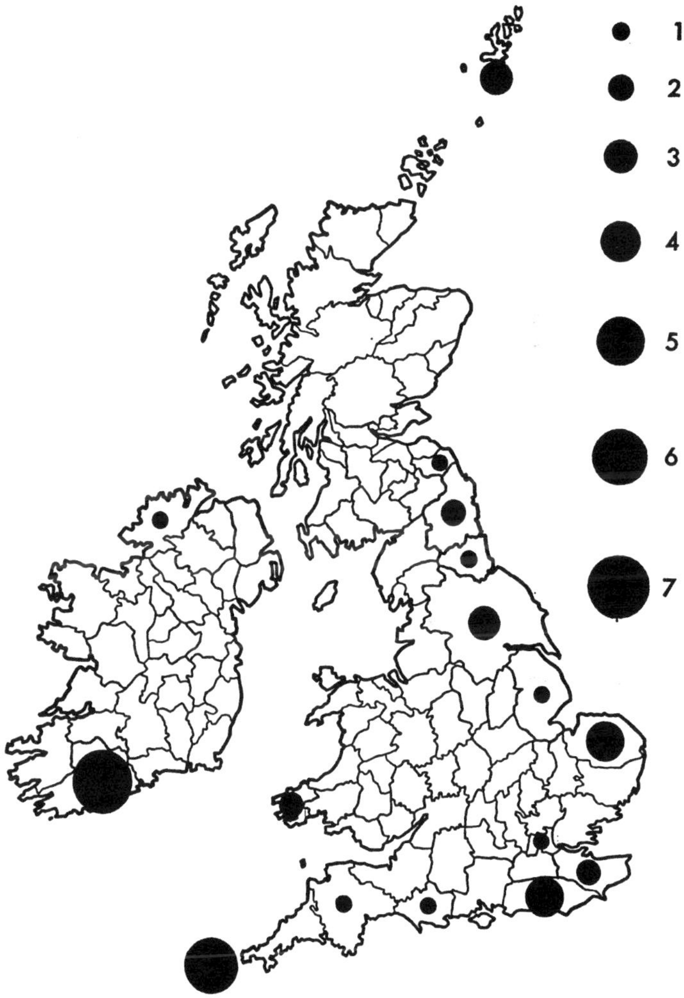

Fig. 81. Distribution by counties of August–November Greenish Warblers *Phylloscopus trochiloides* in Britain and Ireland during 1958–67

1964 to 15th January 1965 in Scilly—and no fewer than three more in December 1968. At least one has also been identified on the Continent in winter, in Switzerland in January 1960 (Roux 1960).

The geographical distribution of the August–November records (Fig. 81) shows a scatter along the British east and south coasts, but the

Fig. 82. European distribution of Greenish Warblers *Phylloscopus trochiloides* with the breeding range of this summer visitor shown in black (reproduced, by permission, from the 1966 edition of the *Field Guide*)

main concentrations were in Norfolk and Sussex (four each), Scilly (six) and Co. Cork (seven). In view of the north-east European breeding distribution (Fig. 82) and eastern primary standard direction towards the Indian wintering area, it is surprising to find that as many as 45% of the autumn records were in western areas (south-west England, Wales and Ireland). Rabøl (1969) has already shown that occurrences in the south-west are later than those in the north and east and this was demonstrated during the ten years by the fact that 66% of those in south-west England, Wales or Ireland occurred in October or November, whereas 73% of those elsewhere were in August or September. The number each autumn (Fig. 83) varied from two to seven and no clear pattern emerges. The ages of only three first-year birds in autumn and an adult in June have been published; only one female was sexed.

In late autumn the problem of identification of Greenish Warblers is complicated by the presence of small numbers of northern and eastern Chiffchaffs *Phylloscopus collybita* of the races *abietinus*, *fulvescens* and *tristis*, which in the field sometimes appear to have small pale wing bars (R. H. Dennis in preparation). The fact that 55% of the

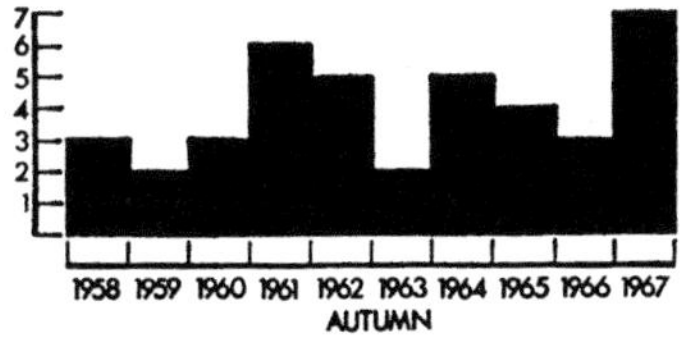

Fig. 83. Annual pattern of August–November Greenish Warblers *Phylloscopus trochiloides* in Britain and Ireland during 1958–67

earlier Greenish Warblers in August–September have been trapped (and, therefore, the identification confirmed in the hand), as against only 17% of the later ones in October–November, has led to speculation whether a number of the latter may have been the result of misidentification of Chiffchaffs with wing bars. This error has certainly occurred on more than one occasion, being rectified after further observation or subsequent trapping, and must clearly be borne in mind. On the other hand, it should not be forgotten that these late individuals have occurred mainly in south-west England and Ireland at localities where recording is traditionally concerned more with observation than with trapping: taking Yellow-browed Warblers in 1958–62 as an example, 42% of those in eastern Britain were trapped, compared with only 17% in south-west England and Ireland. While the higher proportion examined in the hand in eastern Britain must make the identification of Greenish Warblers there more positive, all the records have been assessed by referees aware of the pitfalls and it is probable that the pattern is genuine.

The pattern is, in fact, common to a number of other eastern and southern species which have records distributed on the east coast of Britain in accordance with arrivals from the east and later appearances in the west of Britain and Ireland in circumstances suggesting arrivals from the south or south-east. Rabøl (1969) has concluded that this can be explained by reversed migration in a westerly direction by a part of the population, followed later either by random dispersal from the main areas of arrival in Scotland and eastern England or by redirection of these same individuals along a southerly standard direction. The latter supposes that the birds which have previously been misoriented through 180° (migrating west instead of east) suddenly assume the southerly orientation which is correct for the latter part of their migration and that those in western Britain and Ireland are derived from the northern part of the breeding range. It seems equally if not more likely that the later individuals in the west are from the southern part of the range farther to the east and that, after reversed migration westwards, they have continued, still misoriented through 180°, by turning north instead of south. 'Eastern' vagrants arrive at the western observatories with southerly or south-easterly winds (e.g. Williamson 1960) and on meteorological grounds appear unlikely to be ones which have dispersed after earlier arrival in Scotland and eastern England.

Arctic Warblers have a more northerly breeding range than Greenish Warblers, and Rabøl (1969) has demonstrated that they have a more northerly vagrancy pattern in Britain and Ireland. The 29 in Britain

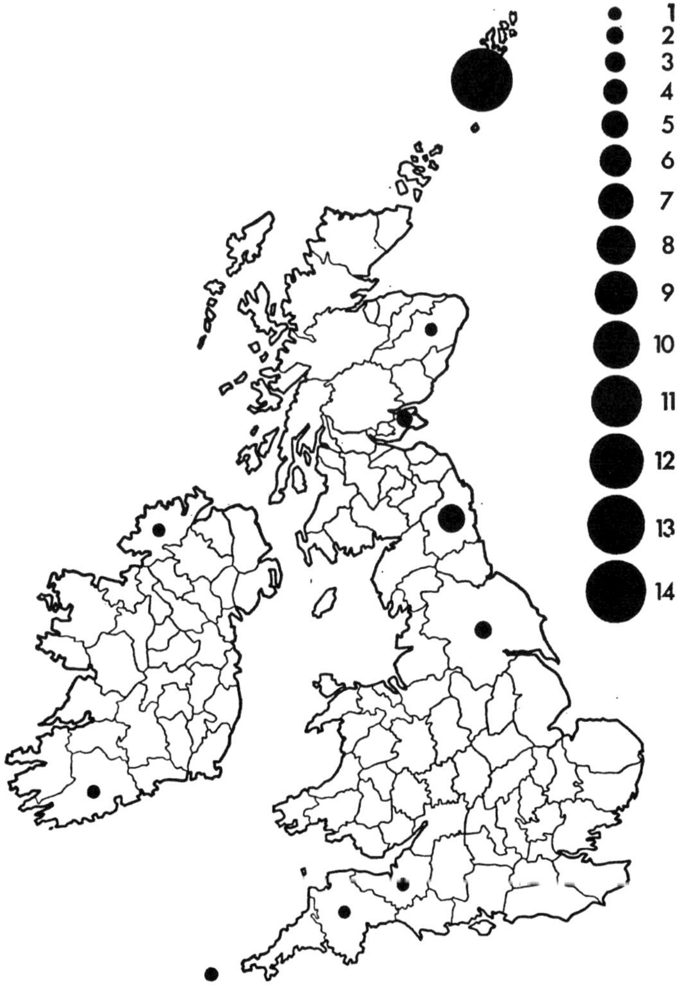

Fig. 84. Distribution by counties of Arctic Warblers *Phylloscopus borealis* in Britain and Ireland during 1958–67

and Ireland in the ten years were all recorded between 14th August and 12th October, with more than three-quarters in the 25 days from 28th August to 21st September, and the annual totals varied from none (1963) to five (1964, 1967). As high a proportion as 58% occurred in Scotland, mainly Shetland (Fig. 84), compared with only 10% of the autumn Greenish Warblers. As well as having a more northerly distribution on the British east coast, the Arctic Warblers also differed in that only 17% were recorded in western Britain and Ireland, compared with 45% of the Greenish Warblers. Neither of Rabøl's two alternative explanations for the late south-western birds takes account of this discrepancy, for both random dispersal and southerly orientation of those in the north should result in a higher percentage in the south-west at a later date, but northerly reversed migration following westerly reversed migration by individuals from the southern part of the breeding range more readily explains the difference. This modification does not alter in essence the important contribution of the main hypothesis put forward by Rabøl (1969).

Scarlet Rosefinch
Carpodacus erythrinus

Unlike Greenish Warblers, Scarlet Rosefinches (formerly known as Scarlet Grosbeaks) have for some time been recognised as irregular autumn visitors, mainly to Fair Isle, Shetland (see *The Handbook*). A total of at least 91 was recorded during 1958–67. Seven were in spring (2nd May to 12th June) and at last 84 (92%) in autumn (mid-August to October); 87% of those in autumn were concentrated in just six weeks from 27th August to 7th October (Fig. 85). The occurrence of seven in spring in ten years is noteworthy as there had previously been only

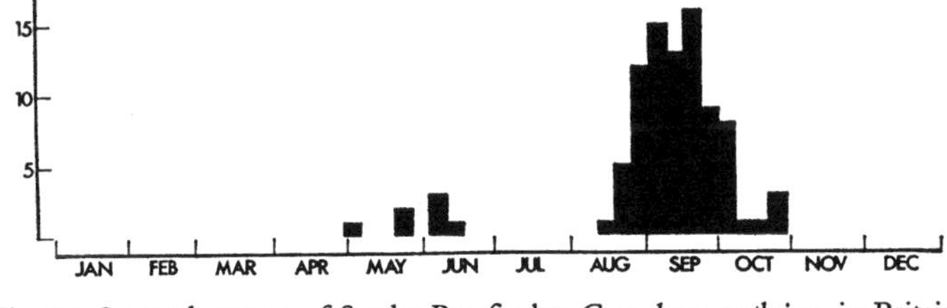

Fig. 85. Seasonal pattern of Scarlet Rosefinches *Carpodacus erythrinus* in Britain and Ireland during 1958–67

one reliable spring record (2nd April 1926 on Fair Isle). Of those in spring, four were in Shetland and one each in Kent, Pembrokeshire and Yorkshire. Two-thirds (at least 56) of the autumn birds were in Shetland, five each in Orkney, Fife and Yorkshire and only one to three in eight other counties (Fig. 87).

The numbers of Scarlet Rosefinches each autumn varied from two to ten during 1958–65 and then there were 16 in 1966 and at least 17 in 1967 (Fig. 86); the spring records were all from 1963 onwards. The recent increases in both spring and autumn may reflect the species' continuing westwards spread in Europe (e.g. Merikallio 1958, Józfik 1960, Turcek 1962–63). The increase in British and Irish records has continued, since there was an average of 22 per year in 1968–72, compared with the nine per year in 1958–67. Ferguson-Lees and Wallace (in Smith 1968), as well as drawing attention to the rise in the number of records, also noted that more than previously were away from Shetland and Orkney and that a higher proportion was now occurring later than the time of the former peak in August and early September. They suggested that Scarlet Rosefinches from populations farther east were now included, though they also considered the possibility that escapes from increasing importations of cage-birds might be obscuring the picture. The escape question must always be borne in mind (see Goodwin 1956, Smith 1967, Scott 1969), but the age pattern of the records seems to militate against this being a serious problem. All but two of those recorded in the ten years were either females or first-year males, which are almost indistinguishable (Williamson 1962), the only two adult males having been on 2nd May 1966 in Kent and on 15th September 1966 in Shetland. This situation—quite different from that of, for example, the Red-headed Bunting *Emberiza bruniceps*, adult males of which greatly outnumber females (Ferguson-Lees 1967)—suggests that most of the records refer to wild birds.

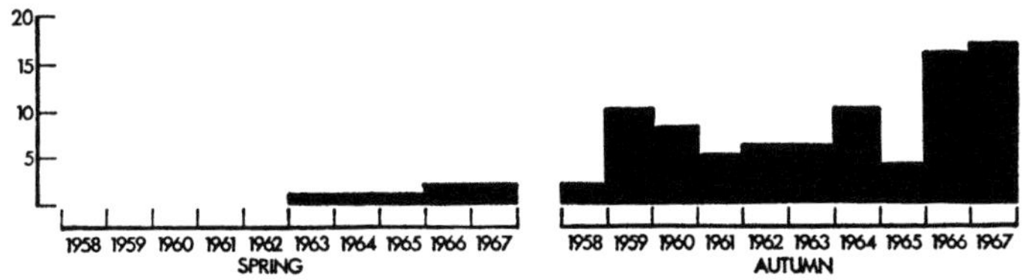

Fig. 86. Annual pattern of Scarlet Rosefinches *Carpodacus erythrinus* in Britain and Ireland during 1958–67 with the spring and autumn records shown separately

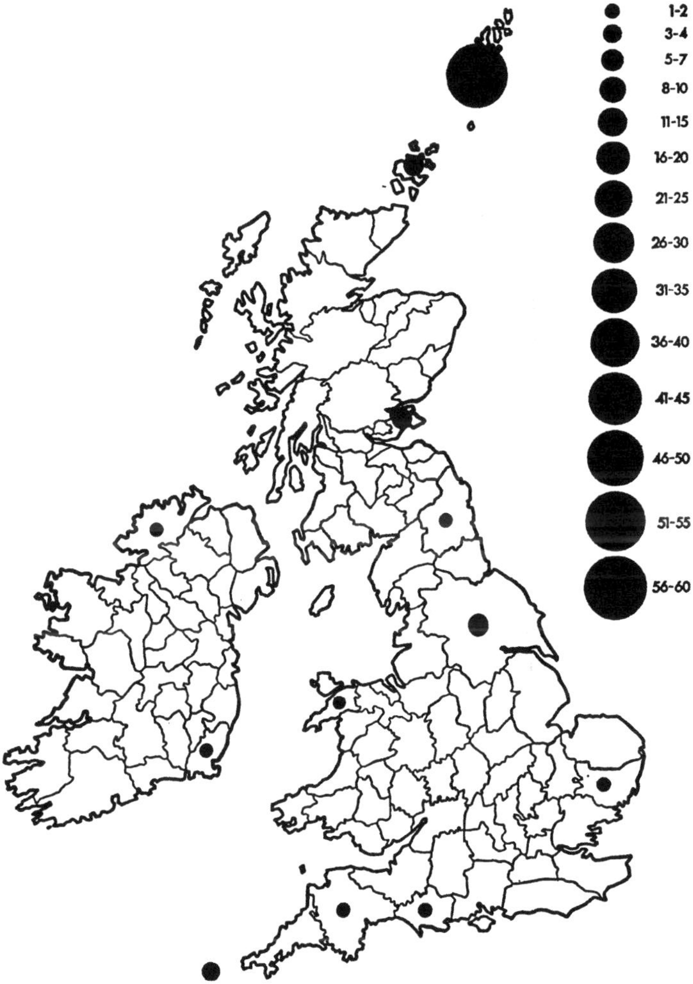

Fig. 87. Distribution by counties of autumn Scarlet Rosefinches *Carpodacus erythrinus* in Britain and Ireland during 1958–67

Even though only small numbers are involved away from the Northern Isles, it is noteworthy that, whereas 80% of the 66 autumn birds in Scotland in the ten years occurred before 17th September, exactly 50% of those on the English east coast (eight) and in south-west England, Wales and Ireland (ten) were recorded after that date. This pattern of later occurrence in the south-west is what one has come to expect, since it is exhibited by almost every vagrant passerine species from the east.

Nisbet (1962) has shown that arrivals of Scarlet Rosefinches at Fair Isle are very strongly associated with low temperatures and northerly winds in Germany and has suggested that the birds reach Fair Isle after flying westwards round the northern sides of depressions centred over southern Scandinavia. Dolnik and Shumakov (1967) found that Scarlet Rosefinches transported eastwards and tested in Kramer cages showed reorientation towards their winter quarters and also reversed reorientation (see discussion by Evans 1968). The European breeding distribution (Fig. 88) is almost the same as that of the Greenish Warbler and their ranges farther east are also similar (Voous 1960); both are more southerly than the Arctic Warbler and yet the distribution pattern of vagrant Scarlet Rosefinches in Britain and Ireland is far more akin to that of Arctic than Greenish. This may be explained by the fact that the wintering area of Scarlet Rosefinches extends as far west as Iran and the standard direction is south-easterly—as against ESE or east at first, followed by south, in the cases of both Greenish and Arctic Warblers, which winter no farther west than India. Reversed migration along a north-west course would be a satisfactory explanation of the extraordinarily high concentration of records in the Northern Isles.

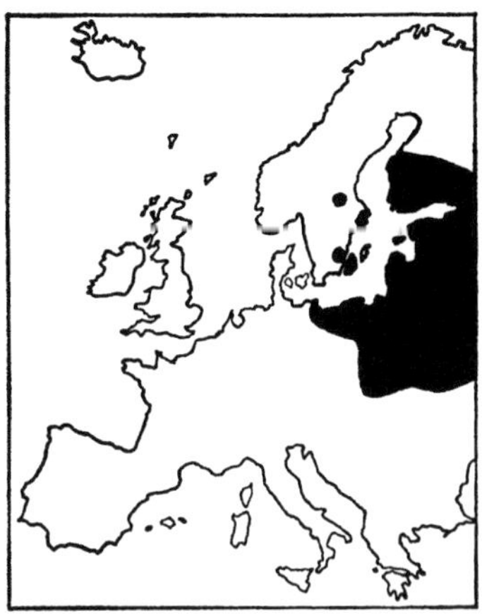

Fig. 88. European distribution of Scarlet Rosefinches *Carpodacus erythrinus* with the breeding range of this summer visitor shown in black (reproduced, by permission, from the 1966 edition of the *Field Guide*)

CHAPTER 7

Mediterranean Gull, White-winged Black Tern and Gull-billed Tern

The breeding ranges of the three species of Laridae considered in this chapter all extend into Europe (Figs. 90, 98 and 104), though are more extensive in Asia.

Mediterranean Gulls are intermediate in size between Black-headed *Larus ridibundus* and Common Gulls *L. canus* and their rather robust shape gives them a jizz closer to the latter even though, in summer, the adults have a black head (not white as in Common Gulls or dark brown like the Black-headed Gull). The adults at all seasons are further distinguished by their white wing tips, with no black. In their various immature and sub-adult plumages, Mediterranean Gulls also resemble Common Gulls more than Black-headed and the best field characters are a heavy, drooping bill and a dark streak extending back from just in front of the eye (not merely a dark spot behind the eye as in winter Black-headed). In Britain they are mainly coastal birds, seldom being seen inland, and are probably best searched for among other gulls at sewage-outlets.

The breeding plumage of adult White-winged Black Terns is very striking and likely to attract attention whenever they occur. The entire body is black, as are the wing-linings, but the forewings, rump and tail are white: giving a very contrasting pattern. The immatures are also readily identified, for they have a dark 'saddle' on the back, clearly cut off by the white nape and rump. In winter plumage there is a greater identification problem and they have to be distinguished carefully from the other marsh terns, Black *Chlidonias niger* and Whiskered *C. hybrida*. The best characters, which need to be observed carefully, are probably the absence of dark 'shoulder' marks, less extensive black on the head than Black Tern, a shorter bill than either of the other species and more leisurely flight. While often seen over the sea, they also occur regularly at freshwater lakes, pools and marshes, dipping to the surface to feed, like other marsh terns (rather than plunging into the water like most sea terns, *Sterna spp.*).

Gull-billed Terns are nearly as large as Sandwich Terns *Sterna sandvicensis* and also resemble them in many ways, being predominently grey and white, with a black cap in summer. The best distinguishing features are their stockier build, shorter and completely black bill (but the yellow bill-tip of Sandwich Terns is sometimes difficult to see), grey rump and tail, and a black patch behind the eye (winter plumage) or streaked buffish crown and nape (immature plumage), rather than the Sandwich's black crown and nape. Unlike most sea terns, Gull-billed have the habit of regularly hawking for flying insects over land. Despite this, most in Britain are seen flying over the sea past headlands.

Mediterranean Gull
Larus melanocephalus

When *The Handbook* was published (1938–41), Mediterranean Gulls were still extreme rarities in Britain (and unknown in Ireland): indeed, only four records are now accepted up to that time, six others having been discarded among the Hastings Rarities (Nicholson and Ferguson-Lees 1962). It is exceedingly difficult to determine the number recorded in Britain and Ireland during 1958–67, since one individual may be seen at several localities or several individuals at one locality and birds at the same place in several years may or may not be the same individual(s) reappearing. Counting one that reappears in several years as a new record on each occasion, but attempting to avoid duplication of wanderers, one can give a conservative estimate of 285 records in the ten years. That this is an arbitrary figure is demonstrated, however, by my calculation of 97 records during 1958–62 compared with 117 estimated from the same raw data by Bourne (1970b). Ages were not published for 32% of the records, but the remainder were 138 adults and 56 immatures or sub-adults. This proportion (approximately 5:2) may not reflect the true situation, however, for adults are easier to detect and identify than immatures and, as shown in table 10, the proportion of immatures was much higher in the second half of the period than in the first. This increase, from less than a fifth to more than a third of those identified, could equally reflect increasing abilities at field identification of young birds or a genuine increase in the number of immatures occurring. Table 10 also shows that almost twice as many Mediterranean Gulls were recorded during 1963–67 as during the previous five years. The annual totals (Fig. 89) increased from only ten in 1958 to a peak of 32 in 1961 and then, after fewer in 1962–63, settled at an average of 40 per year during 1964–67. Once again, however, it is difficult to apportion the effects of increasing observer ability (aided by such papers as Grant & Scott 1967 and Hume & Lansdown 1974), increasing numbers of observers and any increase in the number of birds. The last, however, seems likely to be genuine, since the species has been spreading in recent years. This is perhaps a surprising development, for within the period under review Voous (1960) wrote of the Mediterranean Gull, 'It now possesses only a very limited distribution range with an unmistakable relic character, and is probably in the course of becoming completely extinct'. Until nesting was established in Hungary in the early 1950s, the breeding range was restricted to Greece, Turkey and the north and west shores of the Black Sea. Since then, pairs have nested for the first time in East Germany, Austria, Estonian S.S.R., Netherlands, Belgium and France (many references including, respectively, Rosin and Wagner 1964, Festetics 1959, Aumees and Paakspuu 1963, Commissie voor de Nederlandse Avifauna 1970, Lippens 1970, and Blondel 1966; detailed summary in Taverner 1972) (Fig. 90). In most if not all cases, this 'extralimital' breeding has occurred within large colonies of Black-headed Gulls *L. ridibundus* or

Table 10. Numbers of Mediterranean Gulls *Larus melanocephalus* in Britain and Ireland during 1958–67, also showing separately those for which age details have been published

	1958–62	1963–67	TOTALS
All birds	97	188	285
Adults	57	81	138
Immatures/sub-adults	14	42	56

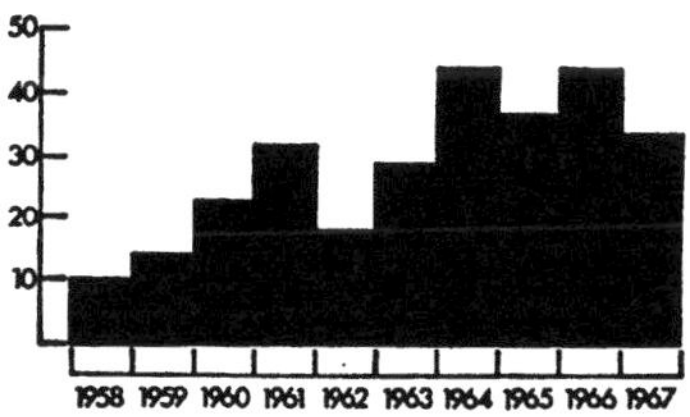

Fig. 89. Annual pattern of Mediterranean Gulls *Larus melanocephalus* in Britain and Ireland during 1958–67

Fig. 90. European distribution of Mediterranean Gulls *Larus melanocephalus* with the normal breeding range shown in black, small breeding populations and spasmodic nesting by open circles, and the main wintering areas by dotted lines (modified, by permission, from the 1966 edition of the *Field Guide*)

mixed colonies of Black-headed and Common Gulls *L. canus*. The increase in British records and these developments on the Continent led Hollom (1957) and Bourne (1970b, but written in early 1968) to prophesy that Mediterranean Gulls might take up summer territories and nest in Britain. In the event, breeding was recorded for the first time in 1968, in a colony of over 10,000 Black-headed Gulls in Hampshire (Taverner 1970).

Although Hampshire is still the only county in which breeding has been recorded, adjoining Sussex produced the largest total of Mediterranean Gulls during 1958–67 with 28%, over twice as many as any other county. Cornwall, Hampshire, Sussex, Kent and Norfolk together accounted for more than two-thirds of the records (Fig. 91).

Just as it is difficult to determine the total during the ten years, so is it difficult to ascertain the pattern of occurrences throughout the year. Fig. 92, however, attempts to show the distribution of records within seven-day periods, each individual being included only within the period when it was first seen. The histogram thus illustrates the distribution of arrival dates and one cannot infer from it the period when most birds were present. The scatter, with at least one first record in every one of the 52 periods, suggests that individuals wander widely and are recorded at several localities; it also suggests that, even though an attempt was made to eliminate duplication, the total of 285 records in 1958–67 is still an overestimate. There is clear evidence, however, of a distinct arrival increasing throughout July and reaching a peak in early August, probably continuing throughout August, September and October, possibly even into November. There is also

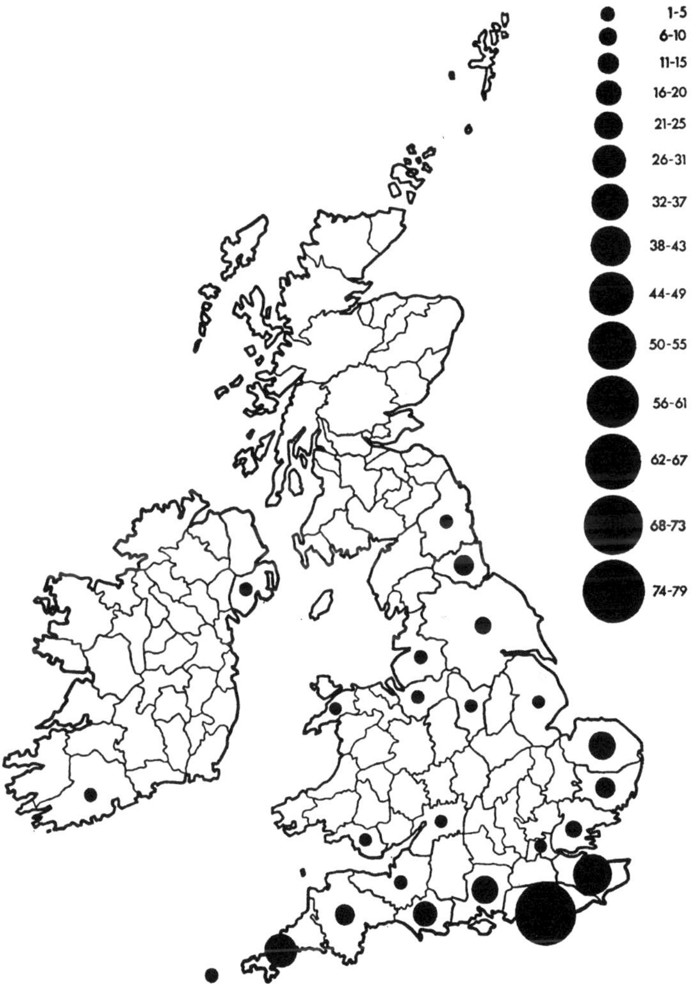

Fig. 91. Distribution by counties of Mediterranean Gulls *Larus melanocephalus* in Britain and Ireland during 1958–67

a smaller peak of new records in mid-April. This pattern was also detected by Bourne (1970b) when analysing the 1958–62 records and he suggested 'that Mediterranean Gulls must often, but not always, undertake three migrations annually: a post-breeding dispersal to

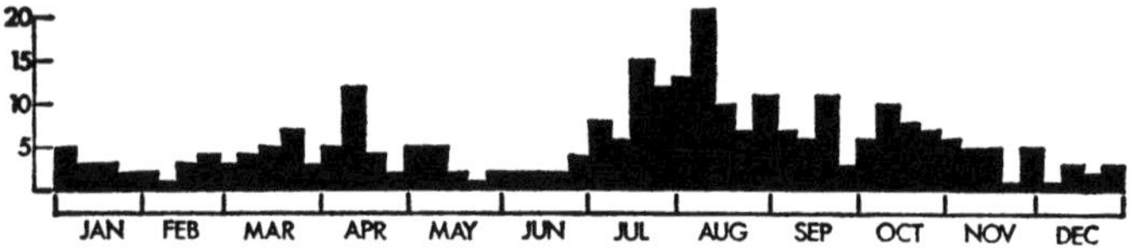

Fig. 92. Seasonal pattern of Mediterranean Gulls *Larus melanocephalus* in Britain and Ireland during 1958–67

regular late-summer quarters where they probably complete the moult . . . followed in some (but not all) cases by a late-autumn movement to regular winter quarters elsewhere before the return to the breeding grounds in the spring.' At all seasons, the largest number of new birds is in south-east England, followed by south-west England and then East Anglia. At the time of spring passage (12th March–22nd April), 58% are in south-east England, a more limited distribution than at any other time of year. Splitting the main autumn period into three 42-day periods (2nd July–12th August, 13th August–23rd September and 1st October–11th November), the proportion in the northern part of England (north-east, north-west and eastern England) is highest in the first period (17%), the proportions in East Anglia and south-west England are highest in the second period (20% and 26%) and the proportion in south-east England is highest in the final period (49%). This suggests that, while there are arrivals and passage in all areas in early autumn, the population becomes stabilised earliest in the north and then in East Anglia and the south-west, and that, by late autumn, wanderers and passage migrants are mostly in south-east England.

No Mediterranean Gulls have been ringed in Britain, but individuals bearing rings have been seen on at least three occasions—in Northumberland in March 1955, in Sussex in July 1956, and in Hampshire in July 1968 (a breeding bird paired with a Black-headed Gull)—but probably other sightings have not been published. It was assumed that the first two of these came from the island of Orlof in the Black Sea breeding area where the fluctuating population has varied between 37,600 pairs in 1936 and 6,000 pairs in 1952 (Schevareva 1955) and where substantial numbers have been ringed. The one in Sussex in July was very early to be so far west, but it is matched by one ringed on Orlof on 2nd July 1949 and recovered less than four weeks later at St Valery-sur-Somme, France, about 25th July 1949. The number of the ring on the one in Hampshire was read through a telescope, however, and it had been marked as a chick at a small colony in Mecklenburg on the

German side of the Baltic. Mayaud (1954, 1956) analysed over 100 recoveries of Mediterranean Gulls ringed on Orlof and showed that the main migration starting in July was westerly, the birds reaching the Adriatic and eastern Libya in August–September and their winter quarters in the Adriatic, south of Italy, Sicily and north Tunisia in September–October, with small numbers in the western Mediterranean and Gulf of Gascony. He also detected a small passage up the Danube and as far west as Switzerland, where a few winter on the lakes. He noted a very small passage from the Baltic to the Bay of Biscay during July–August, returning in March–May, and suggested that these birds had reached the Baltic by travelling NNW along the Dnepr, over the Pripyat marshes and across the Polish steppes. The discovery in the 1950s of small breeding populations in the Baltic and the sighting in Hampshire of a ringed one from this area suggest, however, that some of these birds, including a proportion of those seen in Britain on passage and those which winter in Britain and the Gulf of Gascony, originate from northerly colonies, some of which may still be undiscovered.

Devillers (1964) claimed that this was the only species to follow the Dover–Ostend ships for the whole journey, but suggested that its habits were otherwise similar to those of the Lesser Black-backed Gull *L. fuscus*, which usually remains in inshore waters. In this context, it is noteworthy that the recent increases in Britain have not been matched in Ireland, where the species remains an extreme rarity with only three records in the ten years, the Irish Sea apparently acting as an effective barrier.

White-winged Black Tern
Chlidonias leucopterus

Although the White-winged Black Tern was then considered to be a rare vagrant, *The Handbook* (1938–41) listed a number of records and noted that the species occurred mainly in May, also April and June, but very rarely in autumn (July–November). The preponderance of spring records at that time is hardly surprising, however, for *The Handbook* also noted that in winter plumage the absence of dark markings at the sides of the breast 'affords [the] only means of differentiation' from the Black Tern *C. niger* and that the 'Whiskered Tern [*C. hybrida*] . . . also lacks these patches and in winter plumage this species and White-winged appear to be inseparable in the field.' This attitude was perpetuated in the British literature until the appearance of a paper on the identification of the three *Chlidonias* terns by Williamson (1960a). About 89 White-winged Black Terns were recorded in Britain and Ireland during the ten years, including 31 noted as adults and 31 as immatures. Apart from one in March and three in late October or early November, all were from mid-April to early October, with very distinct peaks during 7th–20th May (13%) and 13th August–2nd September (34%). Splitting the records into those before 2nd July and those after 8th July, 35% occurred in spring and 65% in autumn (Fig. 93). The number each year varied from none to six in spring and one to 13 (1964 and 1967) in autumn (Fig. 94). Although the spring numbers fluctuated with little apparent pattern, those in autumn averaged only just over three per year in 1958–63, but over nine per year in 1964–67 (and this trend continued with an average of almost 18 per autumn in 1968–72). This increase is probably genuine, and not the result of increased observations or higher identification standards, since it did not immediately follow Williamson's paper and was not steady (note the low autumn figure in 1965). The ages of nine of those in

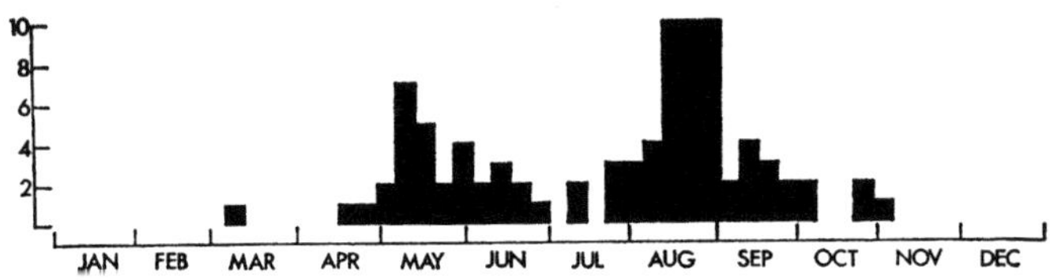

Fig. 93. Seasonal pattern of White-winged Black Terns *Chlidonias leucopterus* in Britain and Ireland during 1958–67

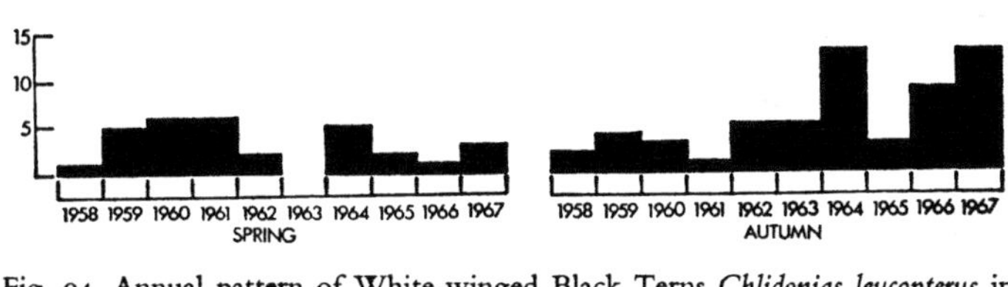

Fig. 94. Annual pattern of White-winged Black Terns *Chlidonias leucopterus* in Britain and Ireland during 1958–67 with the spring and autumn records shown separately

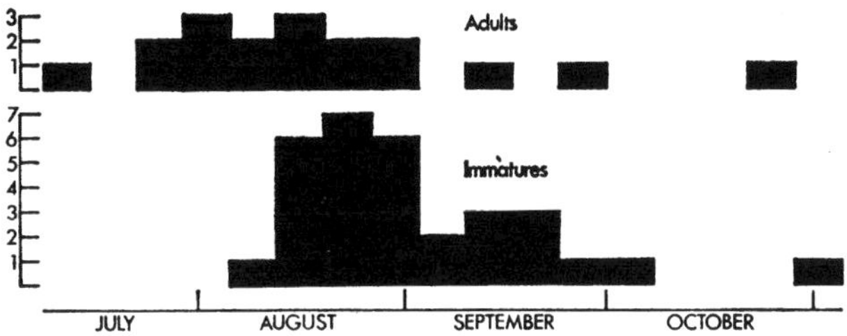

Fig. 95. Seasonal pattern of autumn White-winged Black Terns *Chlidonias leucopterus* in Britain and Ireland during 1958–67 with the adults and immatures/sub-adults shown separately

autumn were not published, but the remaining 49 were 31 immatures and 18 adults. All but three of the adults were in July or August and all but two of the immatures in August or September, this difference being further demonstrated by the fact that 44% of the adults, but only 3% of the immatures, occurred before 13th August (Fig. 95).

White-winged Black Terns were recorded in 20 counties in spring and, while most were in south-eastern England, there was no concentration and no county exceeded the four of Sussex (Fig. 96). In autumn there was almost double the number of records; these were also spread among 20 counties, but they showed more of a concentration in the south-east. Twice as many occurred in Kent as in any other county and Hampshire, Sussex, Kent, Essex, Lincolnshire and London together produced 62% of the records, no other county recording more than three in autumn in the ten years (Fig. 97).

Although they have an almost trans-Palearctic breeding distribution (Voous 1960), White-winged Black Terns nest regularly no nearer to Britain than Hungary and Poland (Fig. 98). The nesting requirements of shallow floods and the varying availability of suitable conditions, depending upon winter snowfall and spring rains, cause very marked fluctuations in breeding numbers from one year to the next (L. Horváth in Bannerman 1962); Wehner (1966) drew attention to this as a possible reason for a recent increase in the number of records in central Europe. He also made comparison with a similar situation in the Black-winged Stilt *Himantopus himantopus*. British and Irish records of the latter, however, do not correlate closely with White-winged Black Terns: for example, the most marked influx of Black-winged Stilts in recent years was 18 in spring 1965, but there were only two White-winged Black Terns then (Harber 1966). In years when there are above average

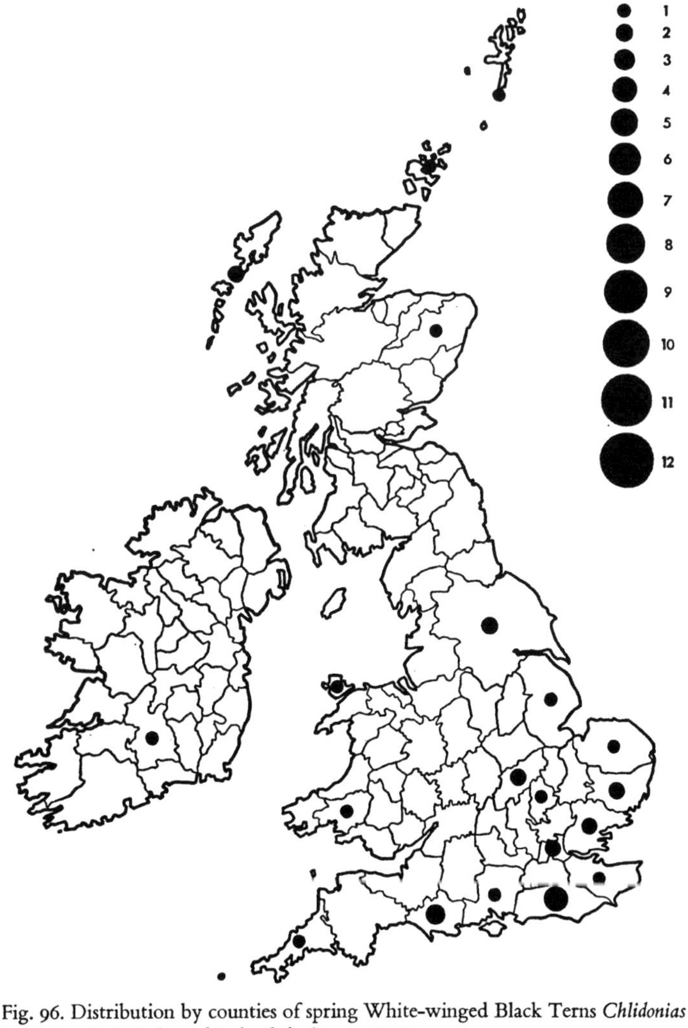

Fig. 96. Distribution by counties of spring White-winged Black Terns *Chlidonias leucopterus* in Britain and Ireland during 1958–67

numbers in spring, one might also expect an above average number of adults and/or a below average number of immatures in autumn, but this pattern is not apparent in the ten years. The lack of correlation in these two instances does not invalidate the argument, of course, since

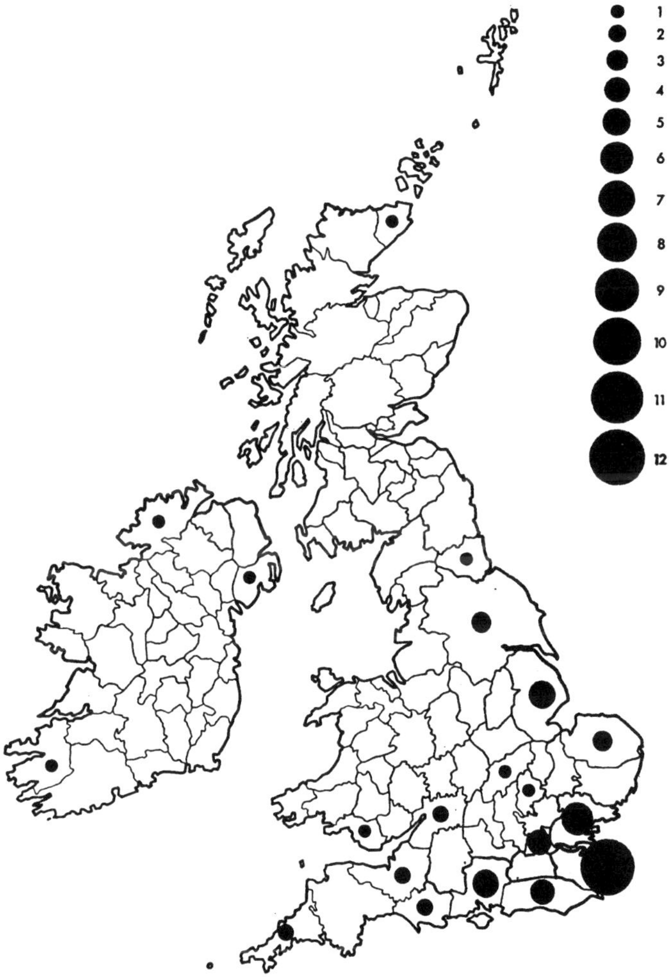

Fig. 97. Distribution by counties of autumn White-winged Black Terns *Chlidonias leucopterus* in Britain and Ireland during 1958–67

different breeding areas could be involved (not only for the two species but also in different years and even seasons). The majority of German (and Swiss) records are in spring, reaching a peak in the second half of May (Wehner 1966); Wehner (1967) and Muller (1967) discussed the

Fig. 98. European distribution of White-winged Black Terns *Chlidonias leucopterus* with the breeding range of this summer visitor shown in black (reproduced, by permission, from the 1966 edition of the *Field Guide*)

likelihood of these birds arriving from the south-west (from west and south-west African non-breeding quarters) or from the south-east (from the larger east African non-breeding quarters). In Britain there has been the assumption that the occurrences in spring of White-winged Black Terns, which arrive concurrently with other 'south-eastern vagrants', and often at the same time as large numbers of Black Terns, in fine anticyclonic conditions with light easterly winds, are the result of overshooting, the birds passing beyond their usual nesting area in conditions ideal for migration. It has been argued by Williamson (e.g. 1960b) that the light winds in such situations are insufficient to cause lateral drift of birds flying north-eastwards from Iberia and that others oriented north-west are involved. The pattern of arrivals in Britain is very similar to that in Hungary, however, with a peak in early May (Wehner 1966 quoting P. Beretzk), earlier than the peak in the intervening parts of central Europe. This suggests that the birds in Britain and Ireland may have been displaced from a north-easterly course or, alternatively, that those in Germany and Switzerland could include some which have overshot earlier and are returning eastwards. Data for Britain and central and eastern Europe for a single year in which there are numerous records in the west would need to be compared to ascertain the full explanation of the apparent discrepancy.

White-winged Black Terns are recorded almost exclusively as spring migrants in central Europe, but the difficulty of distinguishing the winter plumages of the marsh terns has been held responsible for this paucity of autumn records (Wehner 1966). This makes the interpretation of the British and Irish autumn records rather speculative, for there is no picture from elsewhere in Europe with which to compare them. In north Africa, though, an even smaller proportion than in spring is recorded in the west and most pass through Egypt (Bannerman 1962). Assuming, however, that the British and Irish records for which age details do not exist were divided in the same proportions as those for which these data are available it can be calculated that in Britain and Ireland during the ten years there were 31 adults in spring and 21 in autumn. The autumn adults occurred mainly in late July and August. but the breeding population leaves Hungary in the second half of September (Cramp 1968) suggesting that the early autumn birds in Britain and Ireland are failed or non-breeders, perhaps *en route* for winter quarters in western Africa (where only small numbers spend the non-breeding season) after wandering within Europe. The occurrences of immatures, reaching a peak later, distinctly in mid and late August, show a more regular pattern, but it is not clear whether they result from a random dispersal or an oriented migration. It seems possible that after random wandering they are displaced westwards by easterly winds, becoming caught up in movements of west European Black Terns.

It should be borne in mind that these suggestions are based on very few data—an average of only three birds per spring and two adults and four immatures per autumn in the ten years.

Gull-billed Tern
Gelochelidon nilotica

There were about 40 records of Gull-billed Terns up to the time of *The Handbook* (1938–41) and about another 25 in the next twelve years, mostly in May but occasionally in other months during April–September. One pair nested in Essex in 1950 and possibly also in 1949 (Pyman and Wainwright 1952). It is difficult to estimate the number during 1958–67, because of the possibility of duplication. Wallace and Ferguson-Lees (in Smith 1968) calculated that there were 73 records involving over 80 individuals in the ten years; the corresponding figures used in this paper are 67 records involving 86 individuals, indicating that, as with the Mediterranean Gulls, arbitrary decisions have of necessity been taken. The accuracy of these totals is further complicated by the fact that Gull-billed Terns are difficult to identify because of confusion with Sandwich Terns *Sterna sandvicensis*, or perhaps it is more correct to say that observers unfamiliar with Gull-billed may be confused by Sandwich on occasion and that persons assessing records of birds claimed to be Gull-billed always have this possibility in mind. In any event, Wallace (1970) showed that 25–33% of the records of Gull-billed Terns in Britain submitted to the Rarities Committee were rejected: an unknown proportion of these were doubtless genuine.

The 86 individuals in 1958–67 occurred between mid-April and the end of October, with peaks in early May and late August (Fig. 99). There was, however, a considerable number of records in midsummer, between the two peaks, and, taking arbitrary divisions between 27th and 28th May and between 12th and 13th August, there were 25 in spring, 33 in summer and 28 in autumn.

The spring records were very concentrated, with 92% in the contiguous counties of Kent, Sussex and Hampshire and 56% in Sussex alone (Fig. 100). The summer records were rather more widespread, in seven counties, but 73% were in Sussex (14) and Kent (10) while Northumberland (four) was the only other county with more than two (Fig. 101). Gull-billed Terns were recorded in twelve counties in autumn, including every coastal one from Lincolnshire to Cornwall, except Essex. Kent had the highest total (nine), followed by Dorset and Norfolk (four each) and Sussex (three) (Fig. 102). This pattern, with most in Sussex in spring, most in Kent in autumn and these two counties with almost equal numbers in summer, may be due entirely to the coastal configuration, with up-Channel movements more easily observed from the Sussex headlands and down-Channel movements from Kentish stations.

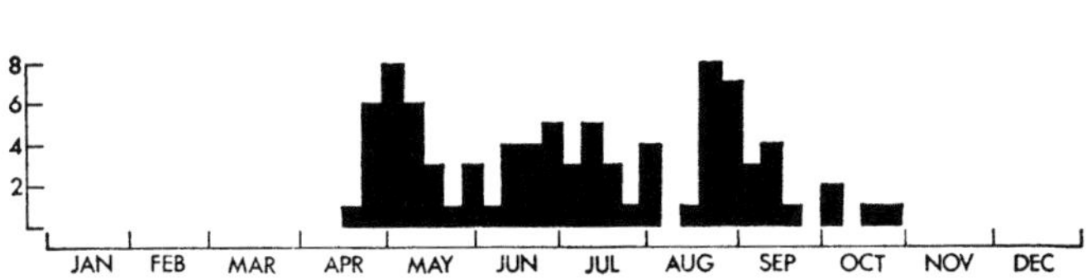

Fig. 99. Seasonal pattern of Gull-billed Terns *Gelochelidon nilotica* in Britain and Ireland during 1958–67

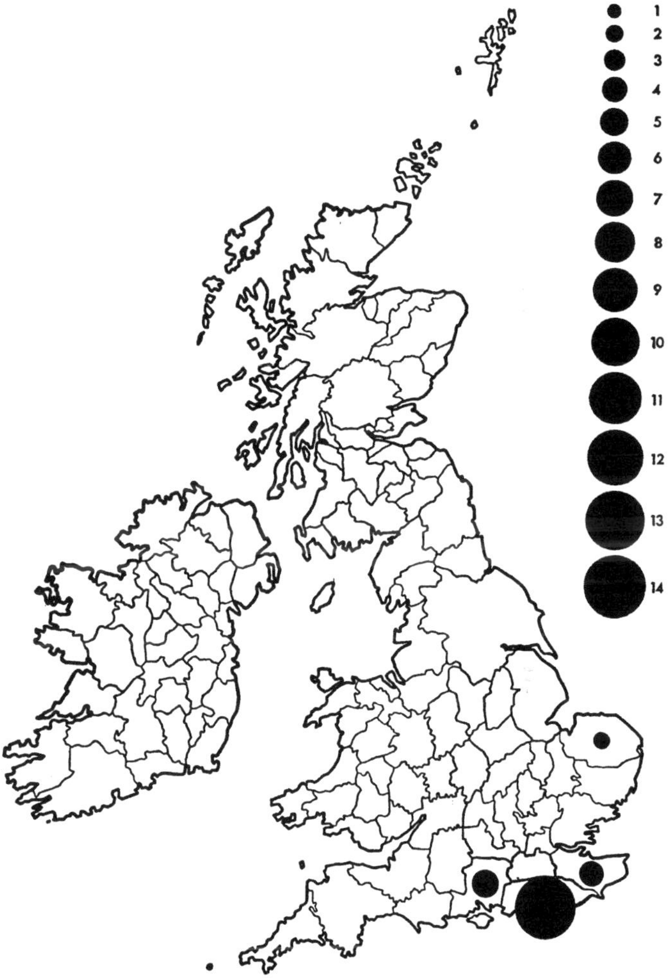

Fig. 100. Distribution by counties of spring Gull-billed Terns *Gelochelidon nilotica* in Britain and Ireland during 1958–67

The highest number in any season was in autumn 1967 (twelve), but otherwise 1960 was remarkable for high numbers at all seasons, with three times the average in spring, four times the average in summer and twice the average in autumn (Fig. 103). It is, however, possible

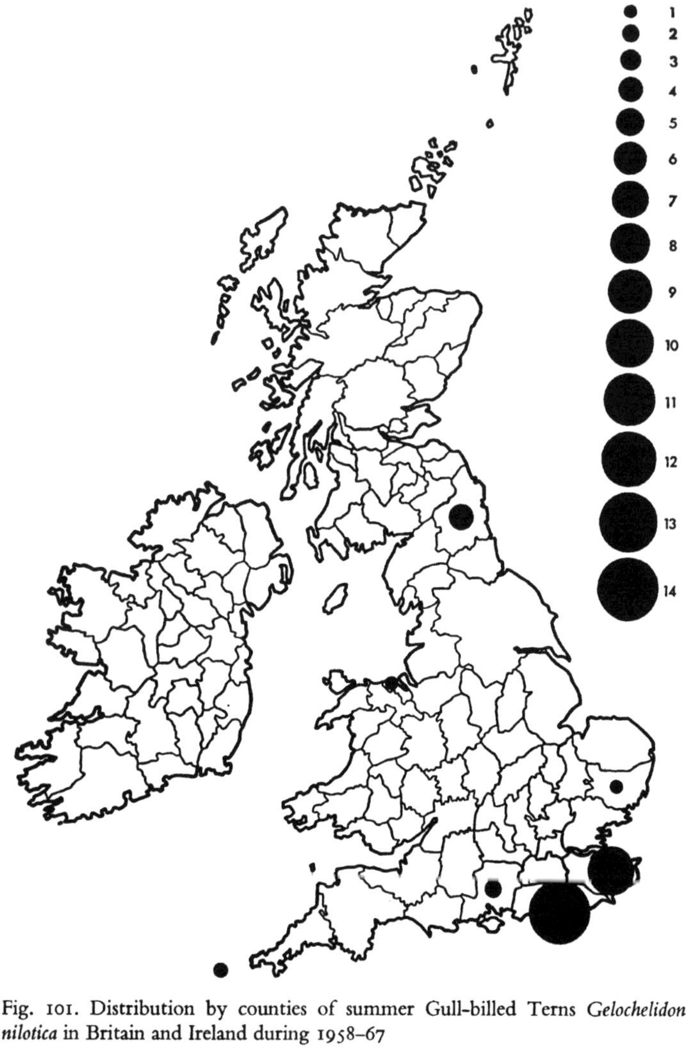

Fig. 101. Distribution by counties of summer Gull-billed Terns *Gelochelidon nilotica* in Britain and Ireland during 1958–67

that many of those 1960 records referred to only a small number of individuals, even though most were seen flying up-Channel (past Selsey Bill, Sussex). The influx in autumn 1967 was undoubtedly genuine, however, without substantial duplication, for six counties

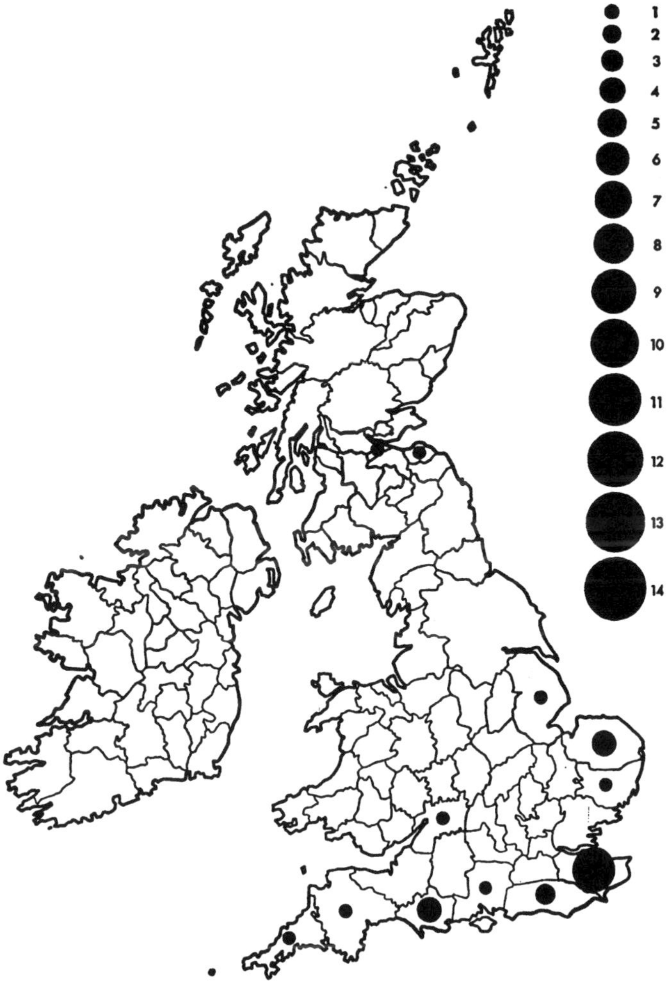

Fig. 102. Distribution by counties of autumn Gull-billed Terns *Gelochelidon nilotica* in Britain and Ireland during 1958–67

were involved and the records were spread over 93 days from 4th August to 25th October (though six were in the eight days from 29th August to 5th September). There has been a decline since this peak, however, with an average of only four per year in 1968–72, compared

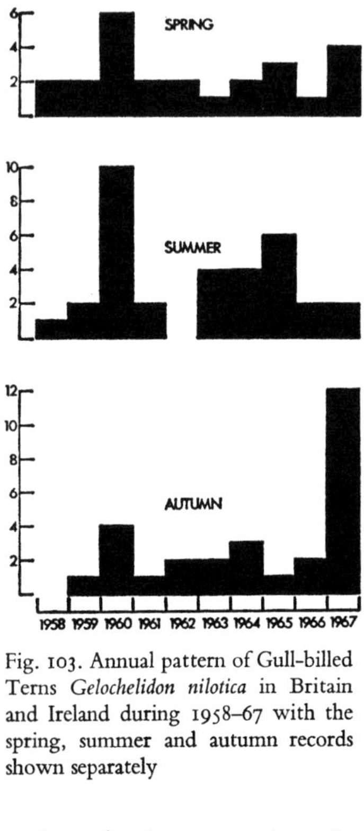

Fig. 103. Annual pattern of Gull-billed Terns *Gelochelidon nilotica* in Britain and Ireland during 1958–67 with the spring, summer and autumn records shown separately

Fig. 104. European distribution of Gull-billed Terns *Gelochelidon nilotica* with the breeding range of this summer visitor shown in black (reproduced, by permission, from the 1966 edition of the *Field Guide*)

with nearly nine per year in 1958–67.

The European breeding distribution includes southern Spain, southern France and Denmark, as well as smaller numbers in Greece and other areas to the east (Fig. 104). Nesting was recorded in the Netherlands in 1931, 1944 and 1945, and then three or four pairs bred regularly from 1949 (Ferguson-Lees 1952) to 1956 and 1958 but not apparently since (Commissie voor de Nederlandse Avifauna 1970). Wallace and Ferguson-Lees (in Smith 1967) and Ferguson-Lees (in Smith 1969) suggested that the Gull-billed Terns seen in Britain are on the way to, or vagrant from, the Danish breeding colonies, rather than overshooting from southern Europe, and the pattern of records, especially the fairly regular occurrences in midsummer, accords with this suggestion. They are perhaps best regarded as regular passage migrants in very small numbers, rather than vagrants.

CHAPTER 8

Yellow-browed Warbler and Richard's Pipit

Both these species have Asiatic breeding distributions but, nevertheless, occur regularly in western Europe; they further have in common a sudden upsurge of records in Britain and Ireland from the late 1960s.

Yellow-browed Warblers are charming little birds, hardly larger than Goldcrests *Regulus regulus*. They are bright olive green above and the underparts are white, faintly tinged with yellow. The best distinguishing marks are two strikingly wide whitish bars bordered with black on each wing (the upper bar shorter than the lower one) and very long, creamy supercilia (hence their vernacular name). The closely allied Pallas's Warbler *Phylloscopus proregulus* is similar but has dark coronal stripes and a broad yellow central crown stripe, yellower wingbars, a striking, square, pale yellow rump patch and habitually hovers while picking food from leaves. Both species are most likely to be seen feeding in the open in trees in this country, but since they usually occur at coastal sites they will take advantage of what vegetation cover may be available.

Richard's Pipits are larger than any other pipit likely to be seen in western Europe, being the same size as Skylarks *Alauda arvensis*. They also resemble Skylarks in many ways, being largely streaked brownish, and they tend to hover prior to landing in much the same way. They are long-legged and long-tailed, but stand in a rather more upright position than Tawny Pipits (the immatures of which otherwise rather resemble Richard's) and have a stronger bill. The call note is very loud and harsh and further serves to distinguish this species. On passage in Britain and Ireland they favour coastal fields, often frequenting areas with longer grass than the short turf preferred by Tawny Pipits.

Yellow-browed Warbler

Phylloscopus inornatus

The Handbook (1938–41) was able to describe the Yellow-browed Warbler as an 'Almost regular passage migrant in very small numbers (more numerous some years than others) along E. coasts Great Britain and in Fair Is. from mid-Sept. to late Oct. . . . Has occurred very rarely in west and once in Ireland.' This statement would still apply today, though, with the Isles of Scilly and parts of southern Ireland now receiving very much greater attention than was the case in the first three decades of this century, Yellow-browed Warblers have been found to occur regularly in the west as well as in the north and east of Britain. A total of 275 was recorded in the ten years, but this included an exceptional influx of about 128 in 1967 and in the other nine years numbers varied from three to 31, averaging 16 per year (Fig. 105). The records were entirely confined to the autumn (mid-September to early December), with 87% in the six weeks from 10th September to 21st October (Fig. 106). *The Handbook* listed five seen in April and May, but none of these is now regarded as fully acceptable (British Ornithologists' Union 1971) and there has been no subsequent spring record in Britain and Ireland.

The species was fairly widespread in the ten years, being noted in 27 counties. All but one of these counties were coastal, however, while 31% of the records were in Shetland and a further 19% in Yorkshire

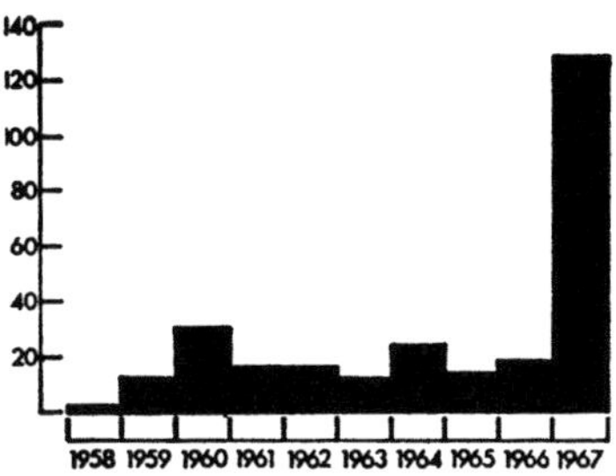

Fig. 105. Annual pattern of Yellow-browed Warblers *Phylloscopus inornatus* in Britain and Ireland during 1958–67 (all autumn)

and Co. Cork. Co. Durham, Isles of Scilly, Kent, Norfolk, Northumberland and Fife, in descending order of frequency, were the only others to average more than one per year (Fig. 107).

Yellow-browed Warblers breed in Asia from the northern Urals east to Anadyr, the Sea of Okhotsk and Ussuriland, northern Mongolia and the mountains of Russian and Chinese Turkestan to north-east Afghanistan and the north-western Himalayas (Vaurie 1959 and Fig. 1 in Rabøl 1969a); they are one of the commonest birds in Siberia (Dementiev and Gladkov 1951–54). The wintering area extends from Afghanistan, India, southern China and Indo-China to the Malay Peninsula. Rudebeck (1956) suggested that a small proportion of the

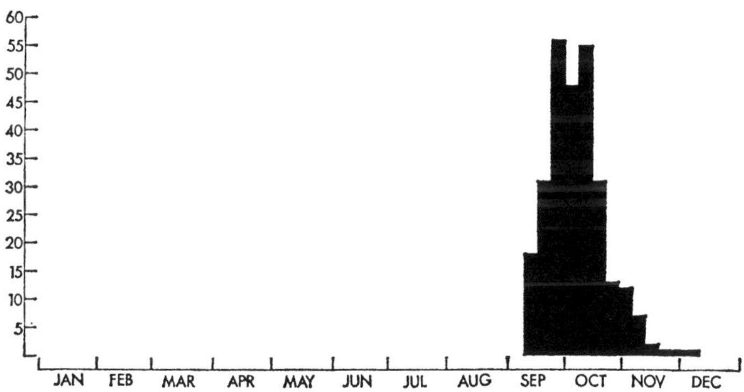

Fig. 106. Seasonal pattern of Yellow-browed Warblers *Phylloscopus inornatus* in Britain and Ireland during 1958–67

population must migrate south-westwards for displacement to occur into western Europe. Nisbet (1962) showed that some of those on Fair Isle had occurred when depressions were situated over southern Scandinavia and postulated that they arrived with north-east winds round the northern sides of such low pressure areas. The westernmost populations mostly lie north of 60°N, however, and the generally northerly distribution of the British and Irish records led Rabøl (1969a), who examined the 1951–67 ones in comparison with those of the more southerly breeding Pallas's Warbler *P. proregulus*, to conclude that the patterns mirrored the breeding ranges: he postulated that the west European occurrences were due to reverse migration in a westward direction by a part of the population.

Previous chapters have shown that there is a strong tendency for occurrences of eastern vagrants to be later in south-west Britain and Ireland than in eastern and northern Britain. This is also true of Yellow-browed Warblers: 56% of the Scottish records were before the begin-

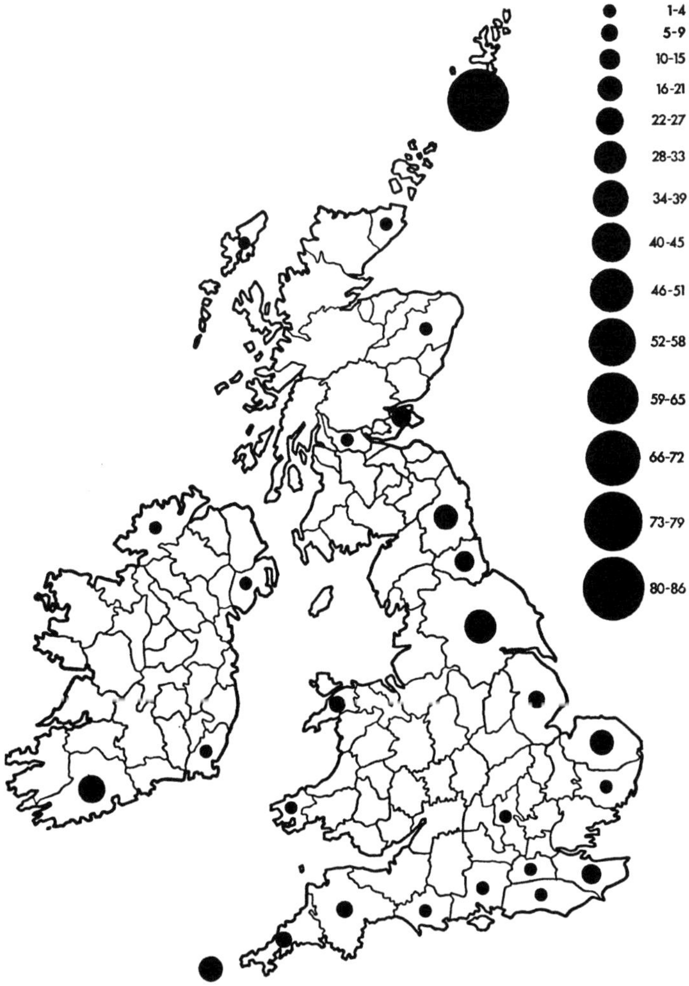

Fig. 107. Distribution by counties of Yellow-browed Warblers *Phylloscopus inornatus* in Britain and Ireland during 1958–67

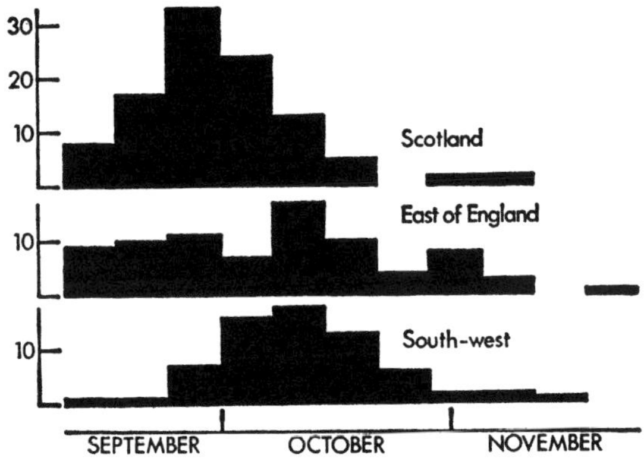

Fig. 108. Seasonal pattern of Yellow-browed Warblers *Phylloscopus inornatus* during 1958–67, with the records in Scotland, the east of England (Northumberland to Suffolk) and the south-west (south-west England, Wales and the south of Ireland) shown separately

ning of October, compared with 38% of those in the east of England from Northumberland south to Suffolk and only 13% of those in the south-west (south-west England, Wales and the south of Ireland) (Fig. 108). The peak in the south-west was a fortnight later than in Scotland and, indeed, the situation is similar to that in the Greenish Warbler (see Chapter 6).

Several of the arrivals of Yellow-browed Warblers in Britain and Ireland have received individual treatment already (Williamson 1959c, 1961c, 1962c, 1963b). The two largest movements deserve some mention here, however, especially since the size of the 1967 one provides sufficient data for analysis of a single year's records, which is bound to give a clearer picture than if those for ten years are amalgamated. All but eight of the 31 in 1960 appeared within a two-week period (Fig. 109 upper) and included simultaneous falls at Fair Isle (Shetland) and Cape Clear Island (Co. Cork) on 27th September, followed by a further arrival at the Irish station only on 30th September and 1st October. The Fair Isle birds came with northerly or north-easterly winds, but both the arrivals at Cape Clear Island coincided with very strong south-easterlies. The situation was entirely different in 1967, when most of the 128 Yellow-browed Warblers were spread over an extended period of seven weeks (Fig. 109 lower); there was none in Ireland and only one in south-west England until four weeks after

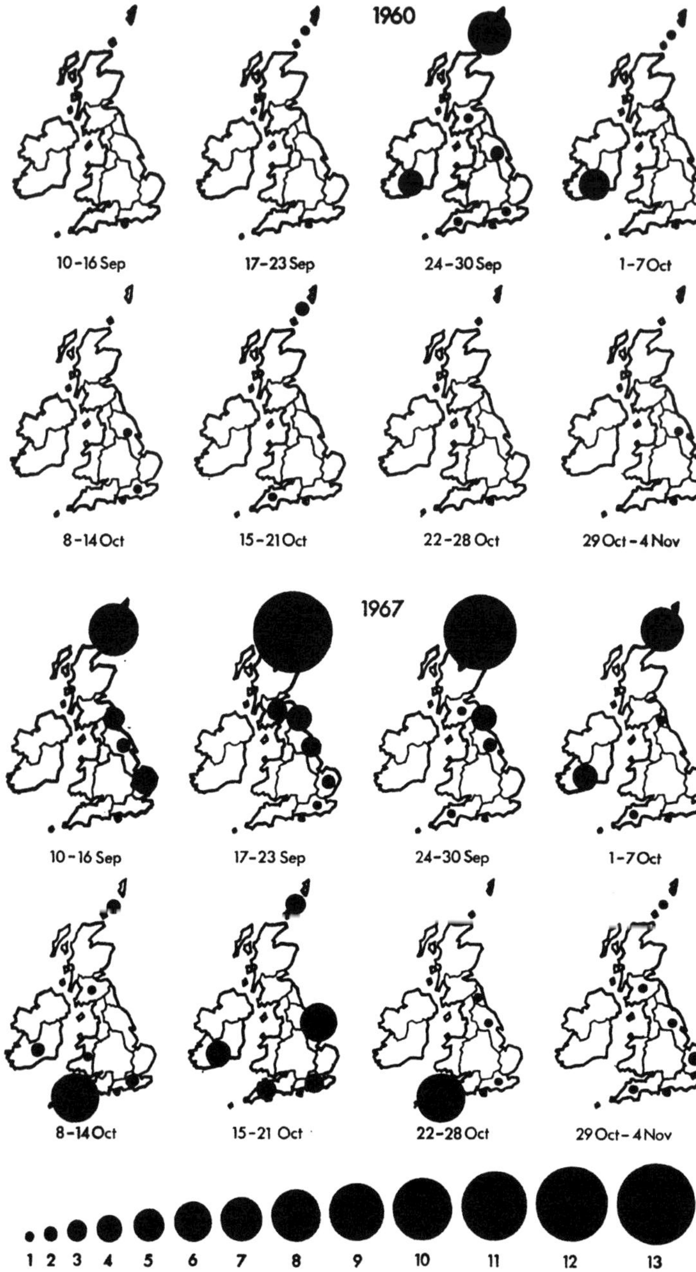
1960
10-16 Sep
17-23 Sep
24-30 Sep
1-7 Oct
8-14 Oct
15-21 Oct
22-28 Oct
29 Oct-4 Nov
1967
10-16 Sep
17-23 Sep
24-30 Sep
1-7 Oct
8-14 Oct
15-21 Oct
22-28 Oct
29 Oct-4 Nov
1 2 3 4 5 6 7 8 9 10 11 12 13

large east coast and Scottish arrivals. Because of the meteorological situation, some of those which appeared later in Ireland were considered to be filtering south after the earlier arrivals in the north and east (Ferguson-Lees and Sharrock 1967), but it should be noted that Cape Clear Island, the only Irish station where Yellow-browed Warblers are regularly observed, was unmanned during 22nd–28th October 1967. The east of England showed a pattern intermediate between those of Scotland and the south west (as in Fig. 108).

This 1967 invasion was also noted elsewhere in Europe. Two caught at separate localities in Belgium on 22nd October (de Cock de Rameyen and Flamand 1968) and three at the Coto Doñana, Spain, on 21st and 31st October and 13th November (Valverde 1968) were the first records for those two countries. In the Netherlands, where only 27 had previously been recorded, there were as many as 34 between 21st September and 13th December, 22 of them in October, and also one extremely early bird on 13th July (Tekke 1968). The records in the Netherlands formed a pattern intermediate between those of the east of England and the south-west of Britain and Ireland: September records made up 73% of the total in Scotland, 51% in the east of England, 26% in the Netherlands and 3% in the south-west of Britain and Ireland.

The records in Britain and Ireland in the ten years (together with those in the Netherlands in 1967) formed a pattern which could be explained in three ways, even after assuming that Rabøl (1969a) was correct in deducing that the major cause of arrivals is reverse migration: these are shown diagrammatically in Fig. 110. First (A), after early arrivals in the north of Scotland, the later peaks in the east of England, the Netherlands and the south-west could have been due to birds filtering south or even turning south on a southern standard direction. Second (B), after early arrivals in the north of Scotland, the later peaks in the east of England and the Netherlands could have been due to birds filtering south and those in the south-west of Britain and Ireland could have been some of these subsequently displaced northwards and north-westwards by south-easterly winds. Third (C), one could postulate that reverse migration occurs on a broad front, but that the southern part of the Yellow-browed Warbler population lags behind

---

Fig. 109 (*opposite*). Regional distribution by seven-day periods of Yellow-browed Warblers *Phylloscopus inornatus* in Britain and Ireland in 1960 (top eight maps) and 1967 (bottom eight). It should be noted that Cape Clear Island, Co. Cork, was unmanned during 22nd–28th October 1967 and, therefore, that the blank for the south of Ireland at that time is not a negative record

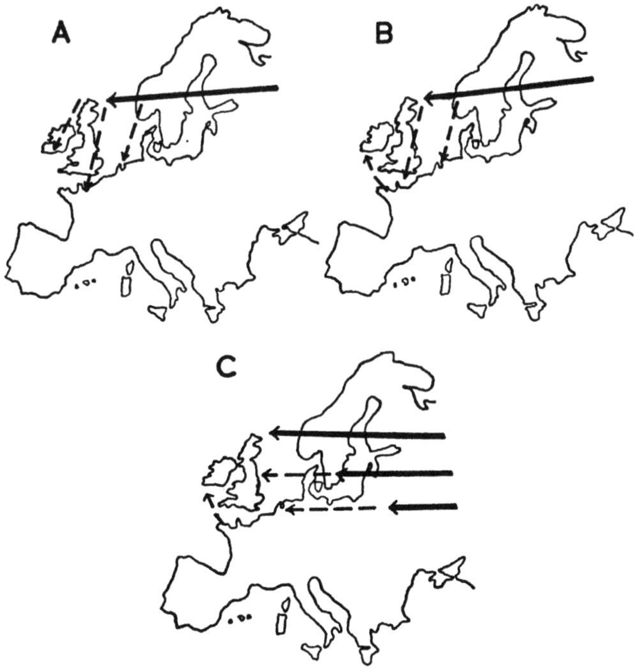

Fig. 110. Diagrammatic representation of three possible explanations of the vagrancy pattern of Yellow-browed Warblers *Phylloscopus inornatus* in western Europe. The thick solid arrows represent movements mostly in September and the thin broken ones those mostly in October: a fuller explanation of A, B and C is given in the text

the northern for one of three reasons. (i) because they start later, (ii) because they have a greater distance to travel or (iii) because the 'centre of gravity' of the population shifts southwards as time progresses (so that those that start later are automatically starting from farther south). In the case of the situation shown in C, those in the south-west of Britain and Ireland could have arrived as a direct part of the main movement, or been displaced by southerly or south-easterly winds from north-west France, or even been following a northerly course on a reoriented reverse migration. Rabøl (1969a) concluded that the south-west tendency in late autumn could point to a random dispersal from the main areas of arrival in Scotland and east England or could be a manifestation of the then southerly standard direction, thus favouring A. In my view, however, all three possibilities probably contribute to the pattern, but C is by far the most important. It is the direct inverse of the situation found in British and Irish records of Hoopoes in spring (and probably of many other species if these were examined in detail), where there are initial arrivals in the south-west followed by others farther along the English south coast and in the south-east, as well as continuing influxes in the south-west, and finally arrivals along the whole east coast, increasingly more northerly as time progresses (see Chapter 1). It seems likely that both patterns reflect the origins of new birds rather than onward passage of the same individuals.

Richard's Pipit
Anthus novaeseelandiae

The Handbook (1938–41) noted that over 100 Richard's Pipits had been recorded up to 1940, chiefly in September–December but also in January–May, mostly in the English south coast counties, in Norfolk and on Fair Isle. At least 212 were recorded during 1958–67, but this total should be regarded with caution because of the difficulty of eliminating duplication. In 1967, for instance, there were 78 bird-days but probably only 15 individuals in the Spurn area of Yorkshire (Fenton and Cudworth 1968) and 64 bird-days but probably only 22 individuals at Fair Isle, Shetland (Dennis 1968). It is even more difficult to estimate the numbers involved at localities which, unlike these, do not have an observatory daily log and a resident warden.

In the ten years there were six records in April–May and the rest extended from the end of August to early January (Fig. 111). Over 72% fell in the six-week period 17th September–28th October, but there were also substantial numbers through November. Since there were several December records but none in February or March, the single bird in January (in Kent on 3rd January 1961) is regarded throughout as a late autumn 1960 record.

Although the average in the ten years was 21 per year, this is a relatively meaningless figure, since there were exceptional influxes in

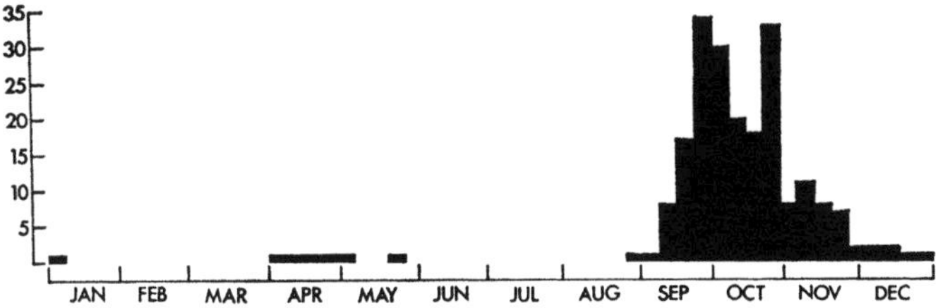

Fig. 111. Seasonal pattern of Richard's Pipits *Anthus novaeseelandiae* in Britain and Ireland during 1958–67

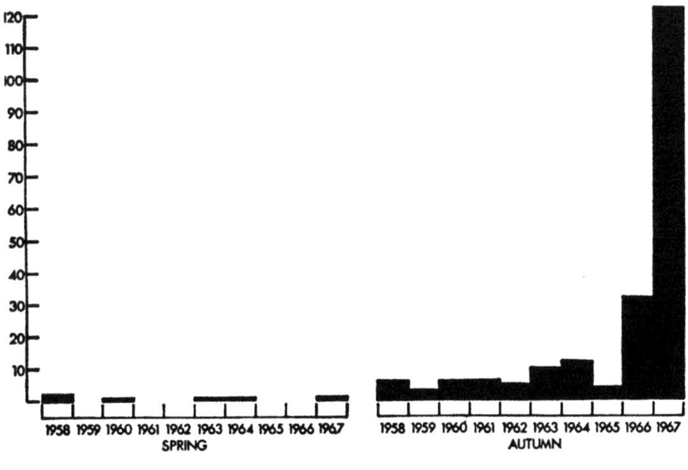

Fig. 112. Annual pattern of Richard's Pipits *Anthus novaeseelandiae* in Britain and Ireland during 1958–67 with the spring and autumn records shown separately

the autumns of 1966, with 32, and 1967, with 122 (Fig. 112); it should be noted that further large influxes occurred in 1968 (130–150), 1969 (over 50) and 1970 (over 100). These recent dramatic events have coloured our view of Richard's Pipits and it may be sobering to reflect that during the eight years 1958–65 there was an average of only six per year, compared with over 100 per year in the four years 1967–70. There is no doubt that many observers have now gained experience which is increasingly enabling them to identify this species with shorter views and even by flight-call alone. Therefore, one may expect recent and future totals to be more realistic than past ones, especially those

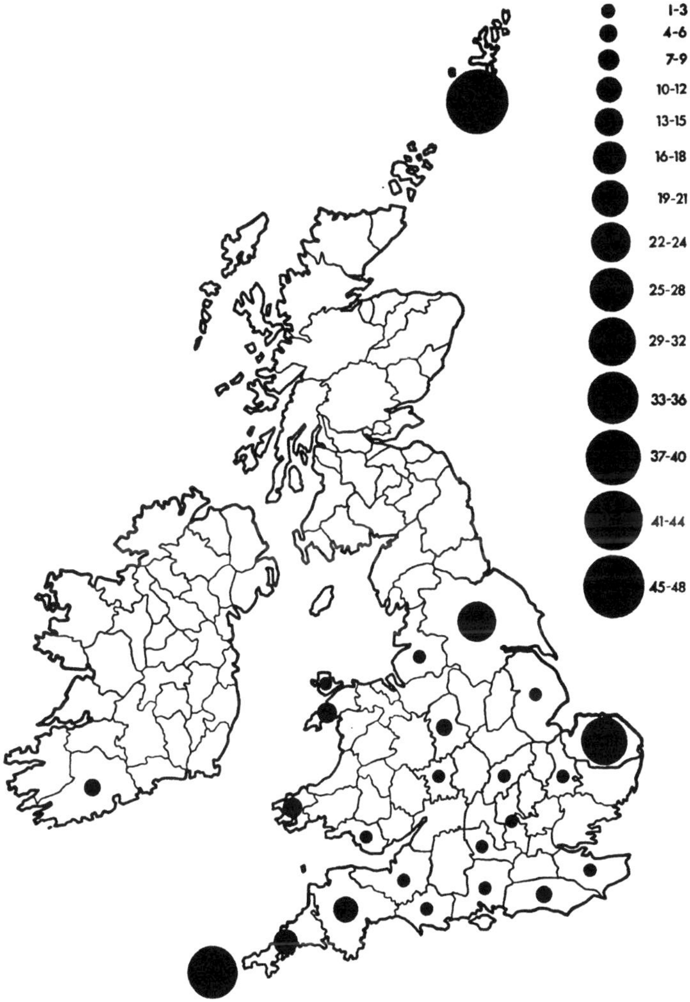

Fig. 113. Distribution by counties of autumn Richard's Pipits *Anthus novaeseelandiae* in Britain and Ireland during 1958–67

before the publication of a guide to the identification of the larger pipits (Williamson 1963d) and subsequent correspondence on the subject (Harber 1964a, Sorensen, Kramshoj and Christensen 1964, Davis 1964), when Richard's Pipits were considered to be easily confusable with Tawny Pipits. Even during the last three years of the ten (1965–67) and subsequently, however, Richard's Pipit has remained something of a problem species and the rejection rate of the records submitted to the Rarities Committee during 1958–67 was in the range 14–24% (Wallace 1970). Recently, Grant (1972) has published a new paper on the identification of Richard's and Tawny Pipits.

In view of the succession of large influxes in recent years, it would be somewhat pointless to analyse in great detail a ten-year period which includes only one of them. At the same time, a full analysis of individual years is not the purpose of this book (see Introduction) and, in fact, the 1967 and subsequent influxes, together with the Continental picture and a discussion of the causes, are to be the subject of a paper being prepared by R. H. Dennis for publication in *British Birds*. Therefore, the treatment here is brief.

The six spring records were in Kent, Surrey, London, Staffordshire, Orkney and Shetland. The autumn pattern was rather peculiar (Fig. 113), with records in 24 counties but two-thirds of them in just four—Shetland, Isles of Scilly, Norfolk and Yorkshire. In England, no fewer than six of the counties were inland, yet four east coast ones (Essex, Suffolk, Durham and Northumberland) had no records. Though Shetland and Yorkshire were two of the four counties with the greatest numbers, not a single Richard's Pipit was recorded in the intervening

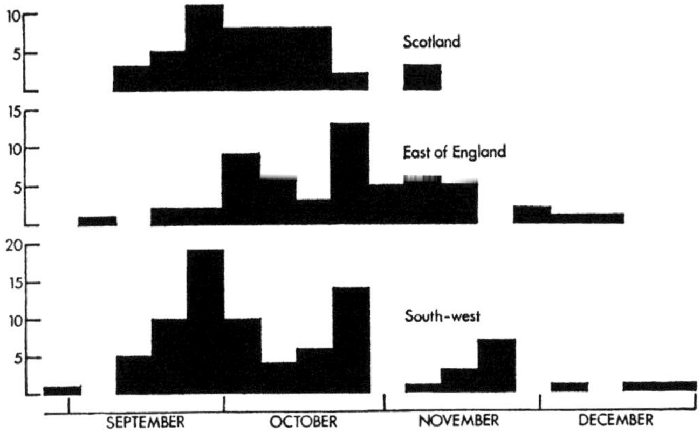

Fig. 114. Seasonal pattern of Richard's Pipits *Anthus novaeseelandiae* during 1958–67, with the records in Scotland, the east of England (Northumberland to Suffolk) and the south-west (south-west England, Wales and the south of Ireland shown separately

area in 1958–67 (and only eight out of over 280 birds in 1968–70—four in Orkney, three in Fife and one in Kincardineshire). Though the Isles of Scilly received the second highest number, only four were recorded in Ireland in the ten years (and there was only one in the next three).

The timing of the arrivals in various parts of Britain and Ireland was also rather strange. In 1966 all but one of those in September were in the south-west (Smith 1967). In the ten years, September records made up 43% in the south-west, 40% in Scotland and yet only 9% in the east of England (Fig. 114) and the pattern was, thus, quite different from that of Yellow-browed Warblers (Fig. 108) and other eastern vagrants. The peaks in the south-west and in Scotland coincided (24th–30th September) and a secondary peak in the east of England came in the following week. The main peak in the east of England (22nd–28th October) coincided with a second peak in the south-west, but was not reflected at all in Scotland. The 1967 movement was also noted in a number of Continental countries, particularly in Finland and the Netherlands. In Finland most of the 59 recorded were during 26th August–1st September (J. Tenovuo *in litt.* to R. H. Dennis) and in the Netherlands there were 47 records involving about 117 individuals from 3rd September to 12th November (Tekke 1968), 51% of them in September. The pattern is unlike that of any other species: after arrivals first in Finland and then in the Netherlands, simultaneous influxes in Shetland and the extreme south-west of Britain and Ireland and, later, arrivals in the east of England and again in the south-west.

The breeding range of *A. n. richardi*, the migratory race of Richard's Pipit which occurs in western Europe, is bounded by the Irtysh in the west, the Yenisei in the east, and the Altai and Tarbagatai mountains in the south, and extends north to 57°N in the west and 59°N in the east (Vaurie 1959). Thus, it lies entirely south of the breeding area of the westernmost Yellow-browed Warblers and immediately west of that of Pallas's Warbler. The wintering area is in south-east Asia and India. Unlike other eastern species, however, the pattern of records in Britain does not reflect this breeding distribution; consequently, post-breeding dispersal, probably with a strong westerly or NNW tendency at first, rather than reverse migration along a firm westerly course, may be the source of the west European occurrences. The situation may also be partly explained by the fact that Richard's Pipit is, to a large degree, a diurnal migrant: it may therefore be more prone to follow coastal guiding lines than other eastern species which are night migrants, as well as being more likely to occur in ideal feeding habitats (many stayed off-passage for considerable periods) and less liable to displacement in adverse conditions. (The British and Irish records of the Tawny Pipit, another largely diurnal migrant, also suggested that displacement in adverse weather had little effect on the pattern—see Chapter 1.) Occurrences in Shetland could then be explained by arrivals from the Norwegian coast of birds which had passed north of or over the northern part of the Gulf of Bothnia, whereas those which had passed south of the Baltic or crossed the Baltic from Finland to southern Sweden may have followed the coast to the Netherlands, northern France and thence south-west Britain, not undertaking a sea crossing until the coastline no longer had a western element. This would help to explain the virtual absence of records between Shetland and Yorkshire, the simultaneous arrivals in northern and south-west Britain and even the later influxes in the east of England (onward passage from north-west Germany and the Netherlands of birds which had stayed off-passage there and then been caught up with flocks of other diurnal migrants). These suggestions are very tentative, but they do provide an explanation of what is otherwise a very puzzling pattern. As already stated, a full study of the several large movements which have now occurred, together with the meteorological situations, is beyond the scope of this study, and they and the reasons for the upsurge in records since 1966 are to be discussed by R. H. Dennis. It is clear, however, that a radical change has occurred: perhaps a westward extension of the breeding range, an alteration in the migratory behaviour of a part of the population, or an explosive population increase. There are no data from the U.S.S.R. to support or refute any of these possibilities.

Finally, the occurrence of six Richard's Pipits in April–May in Britain in the ten years and, for example, three in April and eight in May (out of a total of 23) in the Netherlands in 1960–66 (Tekke 1968) reveals a small but regular spring passage in western Europe and suggests that there may be an undiscovered wintering area in southern Europe or in Africa.

CHAPTER 9

Aquatic Warbler, Barred Warbler and Red-breasted Flycatcher

These last three species all have a mainly east European breeding distribution (Figs. 118, 123 and 129) but have entirely different vagrancy patterns in Britain and Ireland.

Aquatic Warblers are the same size as and closely resemble the much commoner Sedge Warbler *Acrocephalus schoenobaenus*, but have a sandier base colour to the upperparts, which are more heavily streaked (extending on to the rump). The head is boldly striped (even more so than young Sedges, which likewise have a pale central crown stripe) and the tail feathers are pointed—a useful confirmatory feature, when it can be seen. Like Sedge Warblers, they frequent reed beds, and rushes and sedges in marshy areas, but are often even more skulking.

In the adult plumage—seldom seen in Britain and Ireland—Barred Warblers are very distinctive, with greyish upperparts, yellow eyes and whitish underparts with dark crescentic bars. They are large, heavy warblers, larger than House Sparrows *Passer domesticus*, and are rather clumsy in their movements. The immatures show little signs of barring and have dark eyes. They are browner than adults and are best identified by their general greyness, long tails, rather variegated wings and somewhat shrike-like jizz. They are renowned skulkers, but when on passage a patient observer may see them in the open as they feed on berries (ivy and elder, particularly).

The Red-breasted Flycatchers which occur in Britain and Ireland seldom show any red. The orange-red throat, in any case, is a feature only of the adult male. Nevertheless, the species is one of the most delightful birds to occur here. Smaller than all the other European flycatchers, they tend to perch low down on the branches of trees or bushes, from which they frequently drop to the ground to pick up food (but see Davenport 1973), as well as flycatching like the commoner Spotted *Muscicapa striata* and Pied Flycatchers *Ficedula hypoleuca*. Though with mainly mouse-brown upperparts, their dark tails have white patches at the base (like Whinchats *Saxicola rubetra*) and the undertail coverts are also white. They have the habit of cocking their tails upwards, so that these white areas flash conspicuously. A narrow, pale eye-ring is often a useful fieldmark.

Aquatic Warbler
Acrocephalus paludicola

A total of 94 Aquatic Warblers was recorded in Britain and Ireland during 1958–67. Only about 35 had been recorded up to 1940 (*The Handbook*, Nicholson and Ferguson-Lees 1962), and five in 1951 were sufficient to warrant special editorial interest in the journal *British Birds* (Anon 1952c) (a sixth was subsequently recorded that year). The average of more than nine per year in the ten years therefore shows a considerable increase, but the greater number of observers and, especially, the advent of mist-nets are certainly responsible for a large part of this. Mist-nets were not introduced to Britain until 1956 and only about 100 were in operation that year (Spencer 1957), but by 1968 Spencer (1969) estimated that probably about three-quarters of all fully grown birds ringed in Britain and Ireland were caught by that method. No fewer than 45 of the 94 Aquatic Warblers were trapped, probably mostly by means of mist-nets set in reed beds. Table 11 shows the increase in the number of fledged Reed *A. scirpaceus*,

Table 11. Numbers of fledged Reed *Acrocephalus scirpaceus*, **Sedge** *A. schoenobaenus* **and Aquatic Warblers** *A. paludicola* **ringed in Britain and Ireland during 1958–67**

	Reed	Sedge	Aquatic	Reed+Sedge per Aquatic
1958	303	1,181	2	742
1959	771	2,077	9	316
1960	935	1,736	3	890
1961	1,071	1,397	3	823
1962	1,867	2,183	3	1,350
1963	3,083	4,269	4	1,838
1964	3,348	5,542	5	1,778
1965	3,985	8,493	7	1,783
1966	3,264	7,938	5	2,240
1967	6,758	10,365	4	4,281

Sedge *A. schoenobaenus* and Aquatic Warblers ringed in Britain and Ireland during the ten years. The totals of the two commoner species include adults and juveniles ringed in summer and in areas where Aquatic Warblers are unlikely to occur. In addition, the British Trust for Ornithology started an '*Acrocephalus* Enquiry' in 1967 which resulted in 3,494 more fledged Reed Warblers and 2,427 more fledged Sedge Warblers being ringed in that year than in the previous one. Despite these sources of considerable bias, however, the figures show that the Aquatic Warblers ringed during 1958–67 did not increase in step with related species. They do not necessarily prove by themselves that Aquatic Warblers became scarcer (but see below).

There were two records in November and one in May in the ten years, but the other 91 all fell between 2nd August and 9th October, with 81% during the six weeks from 13th August to 23rd September and the peak in the middle fortnight of this period (Fig. 115). The histogram shows a strikingly normal distribution, and D. I. M. Wallace (in Harber 1966) concluded from the smooth distribution curve of the records in 1958–65 that the species should be considered a regular autumn migrant in Britain. His method of analysis was criticised and the smoothness of the curve contested by me (Sharrock 1966), but the data shown here support his conclusion.

The single spring record was in Somerset on 13th May 1963. The two November records (both on 6th, in Devon in 1960 and in Yorkshire in 1967) are grouped with the August–October ones in this paper. The number recorded each year varied from five (1962) to 18 (1959), but it is noteworthy that despite the increase in the number of observers there were fewer in the second half of the period under review than in the first: indeed, the *average* for 1958–62 was exceeded in only one year (1965) during 1963–67 (Fig. 116). This picture of a decline during the

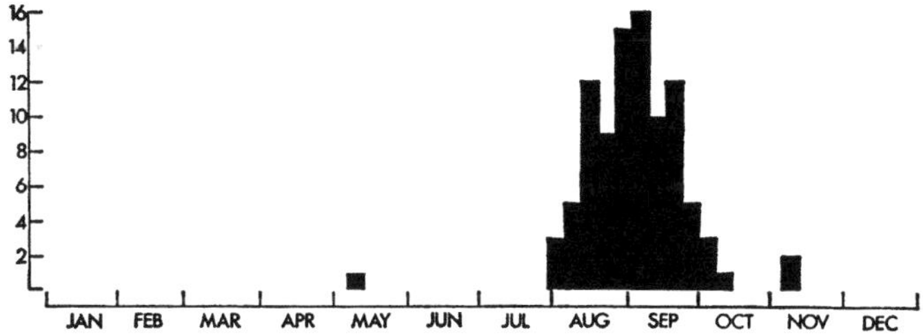

Fig. 115. Seasonal pattern of Aquatic Warblers *Acrocephalus paludicola* in Britain and Ireland during 1958–67

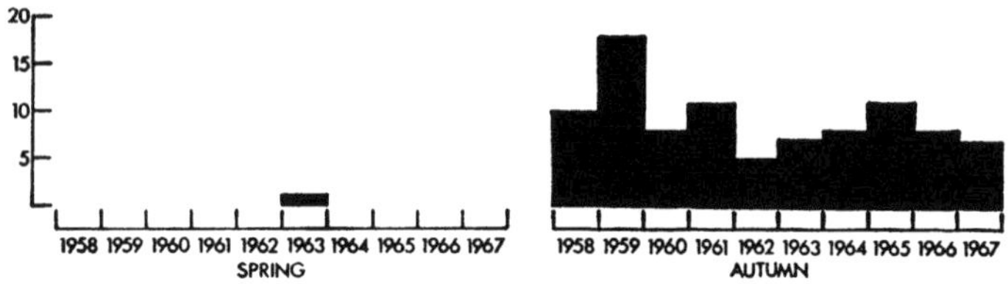

Fig. 116. Annual pattern of Aquatic Warblers *Acrocephalus paludicola* in Britain and Ireland during 1958–67 with the spring and autumn records shown separately

mid-1960s is also implied by the figures in the right-hand column of Table 11. Because of the growing activity of ringers and other observers, it is certain that the actual decrease was greater than Fig. 116 suggests, though there has since been an upsurge of records (outside the ten years under review) with six in 1968, 18 in 1969, eleven in 1970, 27 in 1971 and the astonishing total of 55 in 1972, an average of 23 per year in 1968–72, compared with the nine per year in 1958–67.

Although there were records north to Shetland and west to Co. Cork, the autumn ones were largely confined to the English south coast, with most in Hampshire (Fig. 117). The seven coastal counties from Kent to the Isles of Scilly accounted for 61% of the records and, with the inclusion of the six adjoining counties (mostly inland), the total becomes 74%. D. I. M. Wallace and I. J. Ferguson-Lees (in Smith 1967) pointed out that ringing had shown that Aquatic Warblers occurred with Sedge Warblers in ratios of one in over 1,000 at Chew Valley Lake (inland in Somerset) and one in about 450 at Slapton Ley (on the south coast of Devon) but one in less than 30 on St Agnes (Isles of Scilly). Despite the south coast concentration in England, only four birds were recorded in Ireland in the ten years (where, however, mist-netting is on a small scale). Arrival was almost simultaneous throughout Britain and Ireland: the patterns in south-east and south-west England were identical and arrivals along the British east coast from Norfolk to Shetland and in Ireland all coincided with the peak period on the English south coast.

Since separation of autumn birds into adults and immatures is possible by means of the colour of the upperparts and wing (Williamson 1960c) resulting from the degree of abrasion of the remiges, rectrices and most body feathers (Svensson 1970), it is depressing to find that the ages of only nine of the 45 trapped Aquatic Warblers have been published: all nine were first-year birds.

The European breeding distribution is largely in Poland, Germany,

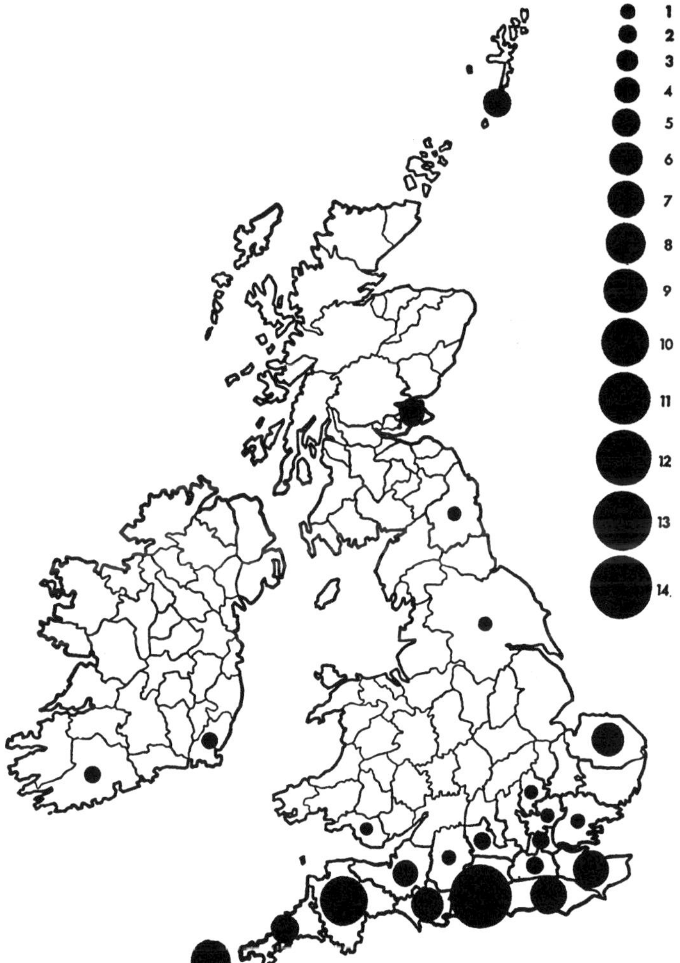

Fig. 117. Distribution by counties of autumn Aquatic Warblers *Acrocephalus paludicola* in Britain and Ireland during 1958–67

Italy, Sicily, Czechoslovakia, Austria, Hungary and Yugoslavia (Fig. 118). This discontinuous distribution was attributed by Voous (1960) to contraction from a previously more continuous range, perhaps partly resulting from the draining of the species' marshland habitat;

numbers have decreased in northern Italy from this cause. Although the breeding range extends to about 60°E, Aquatic Warblers are rare almost everywhere in the U.S.S.R. and a decline has been recorded since the 1920s (Dementiev and Gladkov 1951–54). The wintering area, probably in tropical Africa, is largely unknown, but autumn passage is southerly through the Mediterranean region west to Iberia.

The pattern of autumn records in Britain and Ireland, with simultaneous arrival in all areas and most on the English south coast, but with a proportion penetrating into adjoining inland counties, is entirely consistent with minor displacement of birds migrating south-west from the most north-westerly parts of their European breeding range (Germany, primarily, since Tekke (1973) has pointed out that breeding is not regular in the Netherlands), and bears a similarity to the pattern of occurrences of autumn white-spotted Bluethroats (Fig. 57 and Table 6). If this is the correct interpretation of the pattern, one would expect a proportion of the records to involve adults, and after this conclusion was reached three adults were identified among the 27 in 1971 (Smith 1972). Neither the timing nor the distribution suggests any substantial arrival from the sparse eastern populations. The lack of spring records —only one in the ten years and an all-time total of only four during April–July up to 1968 (British Ornithologists' Union 1971)—is not surprising in view of the presumed northerly, or even NNE, standard direction at that time.

Fig. 118. European distribution of Aquatic Warblers *Acrocephalus paludicola* with the breeding range of this summer visitor shown in black (reproduced, by permission, from the 1966 edition of the *Field Guide*)

Barred Warbler
Sylvia nisoria

At least 516 Barred Warblers were recorded in Britain and Ireland during 1958–67, making this the third commonest species dealt with in this book. Their regular occurrence has been recognised for some time: *The Handbook* (1938–41) noted that they were fairly regular autumn passage migrants in the Northern Isles and probably also annual on the British east coast, with about 50 records in England, many in Scotland, one in Wales and five in Ireland. Baxter and Rintoul (1953) noted that in Scotland there were only two mainland records but many on the islands, and that on Fair Isle up to 40 in a day had been recorded (this on 1st September 1935); during 1948–67, however, the most in a day on Fair Isle was nine (Davis 1964c, Dennis 1967).

Age details were published for only 27% of the records, but it is well known that almost all Barred Warblers in Britain and Ireland are immatures—for instance, no adult had ever been recorded on Fair Isle up to at least 1964 (Davis 1964c)—so that, although the published data show that there were five adults and 136 first-year birds, it is probable that the ratio in the ten years was nearer 1:100 than 1:27, since report editors are likely to specify every adult but make no mention of the age of an immature. As stated in the Introduction, these analyses have used records published in the regional reports with no attempt to assess their validity, on the assumption that they have already been adequately vetted by county recorders. It needs pointing out, therefore, that the records of autumn adults were in Norfolk in 1959, Northumberland in 1959, 1963 and 1964, and East Lothian in 1966, yet Williamson (1964) stated that he knew of no record of an undoubted adult in Britain or Ireland in autumn .

All but two of the 516 records in the ten years fell in the period from August to early November, with 77% in the five weeks from 20th August to 23rd September and an exceedingly pronounced peak (involving nearly half of all the records) in just fourteen days from 27th August to 9th September (Fig. 119). There were two in spring

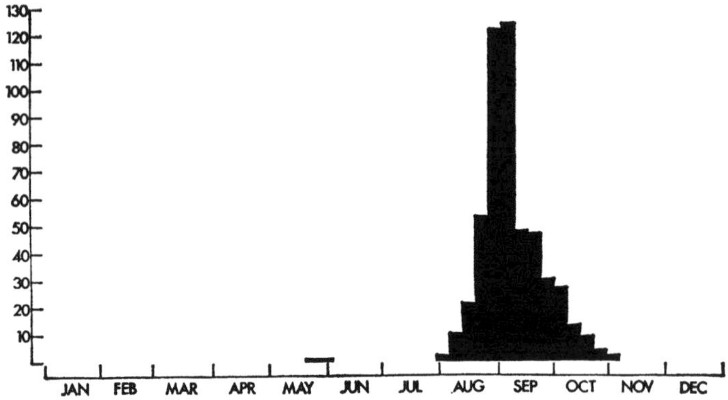

Fig. 119. Seasonal pattern of Barred Warblers *Sylvia nisoria* in Britain and Ireland during 1958–67

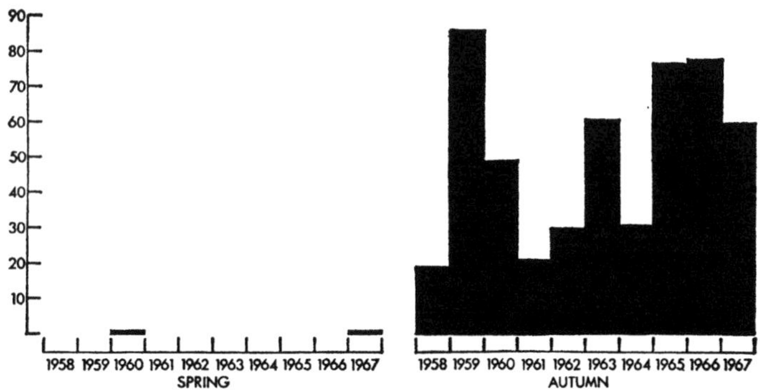

Fig. 120. Annual pattern of Barred Warblers *Sylvia nisoria* in Britain and Ireland during 1958–67 with the spring and autumn records shown separately

in the ten years (Minsmere, Suffolk, 22nd May 1960, and Foula, Shetland, 1st–8th June 1967), the second being an additional record not mentioned in the latest listing of British and Irish birds (British Ornithologists' Union 1971), which notes only one spring record before 1958 (Shetland, June 1914).

Although averaging 51 per year, the numbers of Barred Warblers varied enormously from one autumn to another (between 19 and at least 86), with low totals in 1958 and 1961 and high numbers in 1959,

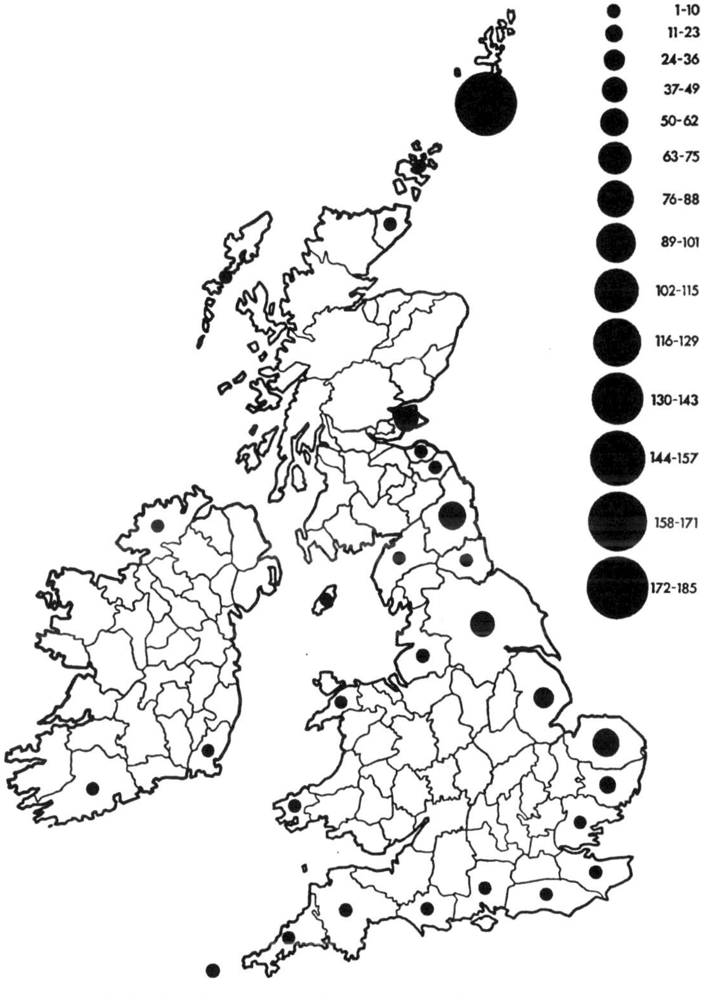

Fig. 121. Distribution by counties of autumn Barred Warblers *Sylvia nisoria* in Britain and Ireland during 1958–67

1965 and 1966 (Fig. 120). Most were in northern Scotland and 36% in Shetland—more than three times as many as in any other county (Fig. 121). The east coast of Britain from Shetland to Norfolk accounted for 83%, the major counties (with an average of between two and six records per year) being Norfolk, Northumberland, Fife, Yorkshire and Lincolnshire, in descending order.

The most northerly records were the earliest, with progressively later peaks towards the south and the latest of all in the south-west (Fig. 122). In Scotland, 27% occurred before 27th August, compared with 8% in the east of England (Northumberland to Suffolk) and 4% in the south-west (south-west England, Wales and the south of Ireland). The peak in the east of England was a week later than in Scotland, though there was no difference between regions within these main groups, the pattern in northern Scotland, for example, being identical to that in southern Scotland. The peak in the south-west came a further two weeks later, 82% of the records occurring after 9th September, compared with 26% of those in Scotland and 35% of those in the east of England. The records in south-east and north-west England are too few for separate analysis, but bear the closest resemblance to those in the east of England. This is a pattern found to a varying degree in many

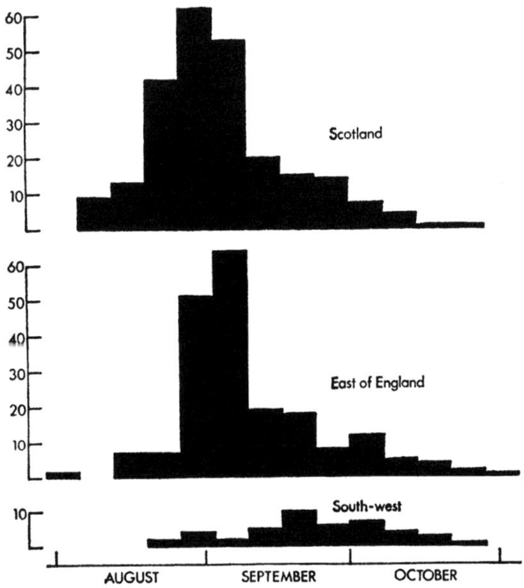

Fig. 122. Seasonal pattern of Barred Warblers *Sylvia nisoria* during 1958–67 with the records in Scotland, the east of England (Northumberland to Suffolk) and the south-west (south-west England, Wales and the south of Ireland) shown separately

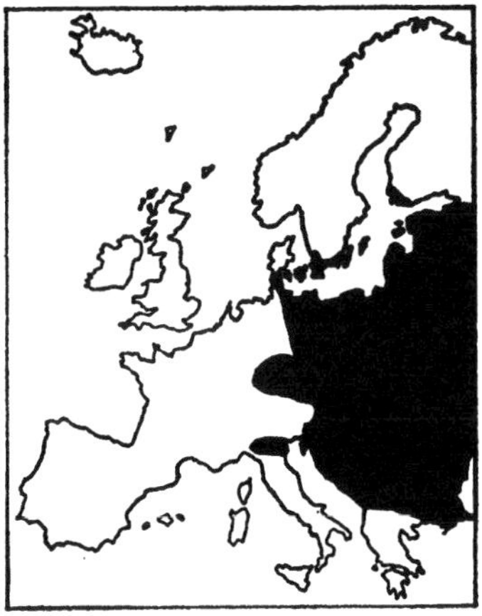

Fig. 123. European distribution of Barred Warblers *Sylvia nisoria* with the breeding range of this summer visitor shown in black (reproduced, by permission, from the 1966 edition of the *Field Guide*)

other species, including Hoopoe, Bluethroat, Icterine Warbler, Greenish Warbler, Yellow-browed Warbler and Scarlet Rosefinch which were dealt with in earlier chapters.

The breeding range in Europe is mostly east of 10°E, from south-east Sweden and Denmark (a small population—see, for example, Preuss 1966), Germany and northern Italy eastwards, between the Baltic States in the north and the Balkans in the south (Fig. 123). In Europe it lies within a band from 40°N to 60°N, but in Asia it is between 36°N and 56°N and extends to 90°E (Voous 1960). Barred Warblers spend the non-breeding season in southern Arabia and east Africa from Uganda south at least to Tanzania (Vaurie 1962, Voous 1960), mostly north of the equator (Davis 1967). The normal migrations of the species are therefore assumed to be oriented NW–SE, and this is confirmed by the absence of records from north-west and west Africa and by its rarity as a migrant in southern Europe west of northern Italy (Williamson 1964). As pointed out by Davis (1967), it is very surprising, in view of the presumed standard direction, that many more Barred Warblers do not occur in Britain in spring by overshooting.

Due to the occurrence of considerable numbers north and west of its breeding range in autumn, this species has attracted a good deal of attention and no little controversy in the literature. The various suggestions may be crudely summarised as follows:

(1) A part of the population migrates south-west instead of south-east, and some of these birds are displaced by easterly or south-easterly winds to Britain (Rudebeck 1956).

(2) There is a random post-juvenile dispersal (Williamson 1959a, 1964, Lack 1960, 1961), especially in years when the population is at a high level, and these birds are displaced by easterly or south-easterly winds to Britain (Williamson 1959a, 1961b, 1963a).

(3) There is a reverse migration oriented north-westwards (Nisbet 1962, Davis 1962), analogous with that of Yellow-breasted Chats *Icteria virens* in the United States (Baird *et al.* 1959), with later arrivals farther south in Britain being the northern birds returning southwards (Davis 1967, Rabøl 1970).

Barred Warblers are rare in continental Europe west of the breeding range—with the first records in Belgium, for instance, not until 1964 (Dambiermont *et al.* 1964)—and relatively scarce in south-east and southern England. All these areas are nearer to the breeding range and involve no sea-crossings or much shorter ones than to the British east coast between East Anglia and the Northern Isles, and this suggests that neither south-westerly migration (1) nor random dispersal (2) can account for the pattern. The data clearly point to reverse migration (3) as the primary cause, though the greatly varying annual totals (Fig. 120) suggest that suitable initiating conditions—'spring-like' anticyclonic weather—at exactly the right time (Fig. 119) are needed, or that the British records reflect the European population level, as suggested by Williamson (1963a). The hypothesis that the birds on the English east coast (in smaller numbers and at later dates than those in Scotland) are on redetermined migration after off-passage stays farther north, however, appears to be contrary to the evidence. On average, they occur one seven-day period later than those in Scotland (Fig. 122), and Davis (1962) noted that in individual years the earliest records at Spurn, Yorkshire, are nearly always later (often by ten days or more) than those on Fair Isle. Yet he himself also noted (Davis 1967) that Barred Warblers seldom remained for more than two or three days on Fair Isle and that long stays were also unusual at Spurn, though more common at the western observatories. Off-passage stays are, therefore, too short to account for the time lags between Scotland and the east coast of England (one week) and the south-west (three weeks). The evidence from recoveries of vagrants ringed in Britain (Table 12) is inconclusive, but only three of the seven—Icterine Warbler, Dusky Warbler *P. fuscatus* and Lesser Grey Shrike *Lanius minor*—are directly relevant and all suggest a failure to reorient after a major displacement. (The lack of recovery data is hardly surprising for, taking one group as an example, the recovery rate of warblers that breed in Britain and

Table 12. Recoveries of scarce and vagrant passerine migrants ringed in Britain

Data from Thomson and Leach 1953, Spencer 1959, 1960, 1964, 1965, 1969, Bonham 1971

	Ringed	Recovered	NOTES
Bluethroat *Luscinia svecica*	7.9.58, Hampshire	9.11.58, Spain	1
Bluethroat *Luscinia svecica*	24.5.59, Fair Isle	28.5.59, Belgium	2
Bluethroat *Luscinia svecica*	21.9.66, Devon	14.9.68, Devon	3
Icterine Warbler *Hippolais icterina*	2.9.64, Isles of Scilly	3.9.64, at sea 137 km WNW	4
Dusky Warbler *Phylloscopus fuscatus*	14.5.70, Isle of Man	5.12.70, Co. Limerick	5
Lesser Grey Shrike *Lanius minor*	13.9.52, Northumberland	15.10.52, Aberdeen	6
Rustic Bunting *Emberiza rustica*	12.6.63, Fair Isle	15.10.63, Greece	7

1. Suggesting reorientation after minor (or nil) displacement, presumably from Scandinavia.
2. Suggesting rapid reorientation after overshooting in spring.
3. Retrapped at ringing locality, suggesting regular passage.
4. Continuing movement away from wintering area, suggesting failure to reorient.
5. Suggesting failure to reorient after arrival in western Europe, possibly in autumn 1969.
6. Suggesting failure to reorient in autumn, or long off-passage stay and continuing northwards movement.
7. Suggesting successful reorientation after spring displacement, perhaps minor.

Ireland is only 1 in 250 and of those which occur only as migrants or vagrants less than 1 in 400, while up to 1969 only 495 Barred Warblers had been ringed in these islands; despite this, one might have expected one or two recoveries by now if Davis and Rabøl were correct in thinking that the English east coast birds are the result of a regular passage from the Northern Isles.)

The facts point clearly to the hypothesis that occurrences in *all* areas result from reverse migration along a north-westerly course from an area of origin which moves south (and east) as the autumn progresses and, therefore, produces smaller and smaller numbers. This is essentially the same pattern as in Yellow-browed Warblers

(Fig. 110), though the direction of the reverse migration is north-west instead of west. The German Barred Warbler population departs in August (Niethammer 1937), and evacuation south-eastwards of most of the birds at this time adequately explains the paucity of records in Britain and Ireland after mid-September, even in the south-west. It seems likely that a high proportion of these north-westerly oriented autumn juveniles perishes in the Atlantic after overflying Britain. There seems little justification for assuming that most make landfall in Shetland and even less for assuming that those that do are able to appreciate that they must promptly turn in a non-suicidal southerly direction.

Red-breasted Flycatcher
Ficedula parva

A total of 496 Red-breasted Flycatchers was recorded in Britain and Ireland during 1958–67. This average of 50 per year compares with a grand total of 85 or more in *The Handbook*, and of fewer than 112 during the 100 years up to 1957 calculated by Williamson (1962c).

The majority were in autumn, with 96% occurring between mid-August and mid-November, the peak during the first seven days of October, and 68% of the autumn birds in the four weeks from 17th September to 14th October (Fig. 124). There were, nevertheless, 19 spring records, from late April to early July. Williamson (1959b) suggested a possible connection between three in spring 1959, which equalled the previous spring total (Radford 1968), and the large arrivals of Red-breasted Flycatchers in autumn 1958, considered at the time to be the outstanding feature of that autumn (Williamson 1958); but there were up to four spring records annually during 1959–65, though none in the next two years. In autumn, numbers were remarkably steady in the ten years, with annual totals of 40–52 in every year except three, these being 1959 (66), 1963 (34) and 1965 (57) (Fig. 125); 1959 was also the best autumn for at least 15 years for Red-breasted Flycatchers on Heligoland (Feeny 1959).

The annual totals shown in Fig. 125 differ from those given by

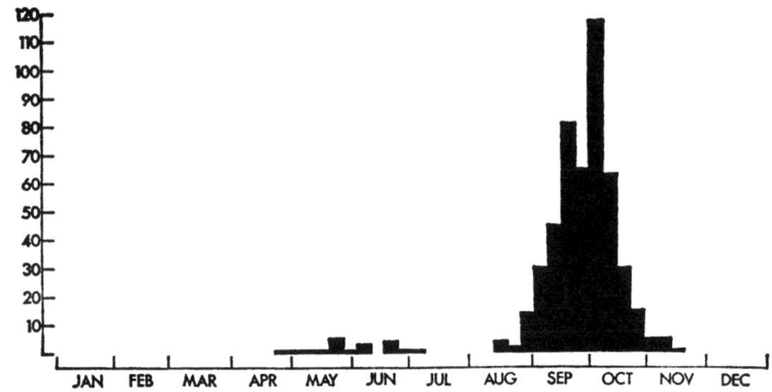

Fig. 124. Seasonal pattern of Red-breasted Flycatchers *Ficedula parva* in Britain and Ireland during 1958–67

Williamson (1959a, c, 1961c, 1962c) which were compiled rapidly at the end of each migration season, mainly from observatory records: in some years additional records were reported some time later and in others reassessment by local recorders made allowance for duplication; also the figures in Williamson (1961c) of about 30 in 1956 and about 60 in 1958 should in fact refer to 1958 and 1959 (K. Williamson *in litt.*).

Of the 496 birds recorded in the ten years, the age was published for only 13% and the sex for 9%: the totals are shown in Table 13. Even in spring some second-year males are inseparable from females (Svensson 1970); the character upon which ageing depends in autumn (presence

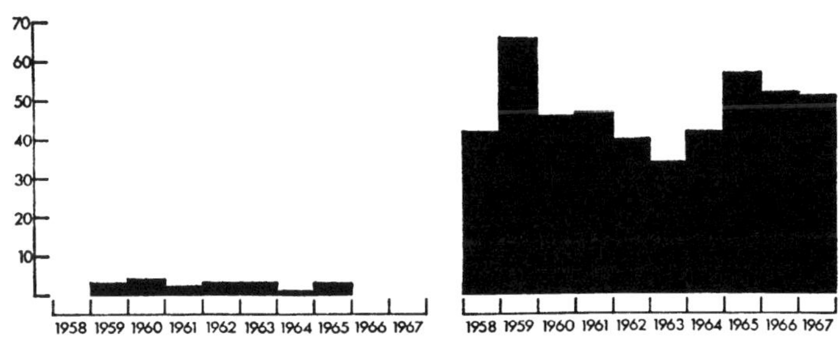

Fig. 125. Annual pattern of Red-breasted Flycatchers *Ficedula parva* in Britain and Ireland during 1958–67 with the spring and autumn records shown separately

Table 13. Published sex and age data for Red-breasted Flycatchers *Ficedula parva* in Britain and Ireland during 1958–67

	SPRING	AUTUMN		
	Whole period	Whole period	Before 17th Sept	On and after 17th Sept
Adult ♂♂	4	14	8	6
Adult ♀♀	2	3	1	2
Unsexed adults	0	1	0	1
Immature ♂♂	0	3	1	2
Immature ♀♀	1	3	0	3
Unsexed immatures	0	32	4	28
Unaged ♂♂	3	2	0	2
Unaged ♀♀	4	6	1	5
Unaged and unsexed	5	413	81	332
TOTALS	19	477	96	381

or absence of yellow-buff tips to, especially, the greater coverts) may not be easy to check in the field and is not of proven reliability. It is not surprising, therefore, that the majority of autumn birds were both unaged and unsexed. The criteria used in determining the ages and sexes shown in this table are not known, the data published in regional reports being taken at their face value, and a number of features in the table cast doubt upon some of the sex/age identifications. Those least likely to include errors are the adult males and it is clear that in autumn adults generally arrive in Britain earlier than immatures: 57% of the adult males (and 50% of all adults) occurred before 17th September compared with only 13% of the immatures (20% of all birds).

The spring birds were in nine counties, but more than two were recorded only in Yorkshire (four) and Shetland (five) (Fig. 126). The autumn ones were in 27 counties, all but one coastal, but seven of these (Yorkshire, Norfolk, Scilly, Shetland, Northumberland, Fife and Co. Cork, in descending order) each averaged two to six birds per year and accounted for 70% of all the autumn records (Fig. 127). This distribution—roughly equal numbers in five east coast counties from Shetland to Norfolk and in the Isles of Scilly—is strikingly different from that of Barred Warblers (Fig. 121), which occurred principally in Shetland. The timing was also different, with arrival almost synchronous in all areas, but with peaks in Scotland and the east of England (Northumberland to Suffolk) in mid-September and in the east of England and the south-west (south-west England, Wales and the south of Ireland) in early October (Fig. 128). In the east of England 29% occurred before 17th September, compared with 16% in Scotland and 12% in the south-west; while in the south-west 64% occurred in October and early November, compared with 42% in Scotland and 41% in the east of England. The early east of England movements were particularly apparent in Yorkshire, Lincolnshire and East Anglia, while the late south-western ones were especially evident in south-west England. The small numbers in south-east England showed a pattern most similar to those in south-west England. The differences in the timing are relatively slight, but it is clear that in the ten years the main arrivals occurred first on the English east coast, followed by later ones in Scotland and the last in the south-west.

The European breeding range of the Red-breasted Flycatcher (Fig. 129) is very similar to that of the Barred Warbler (Fig. 123), extending from Sweden and Denmark (both small populations), southern Finland,

Fig. 126. Distribution by counties of spring Red-breasted Flycatchers *Ficedula parva* in Britain and Ireland during 1958–67

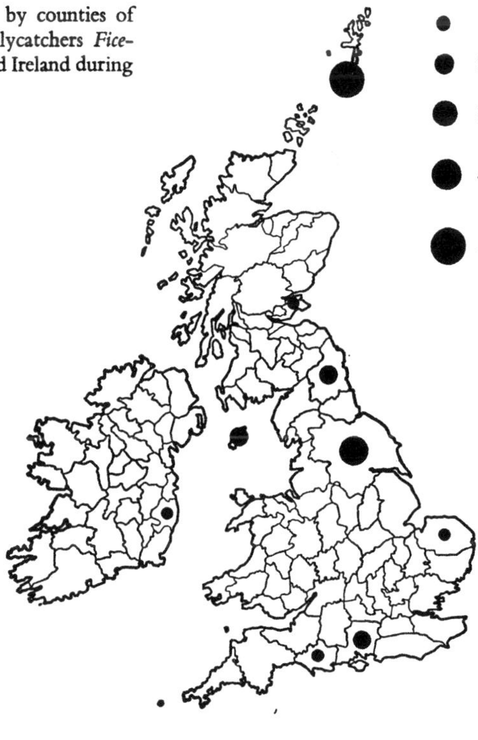

Germany, Austria, Hungary, Yugoslavia, Bulgaria and Romania eastwards, mostly between 40°N and 63°N. Farther east, however, the two distributions differ, since Red-breasted Flycatchers extend to 175°E beyond Kamchatka and their range is more northerly, mostly

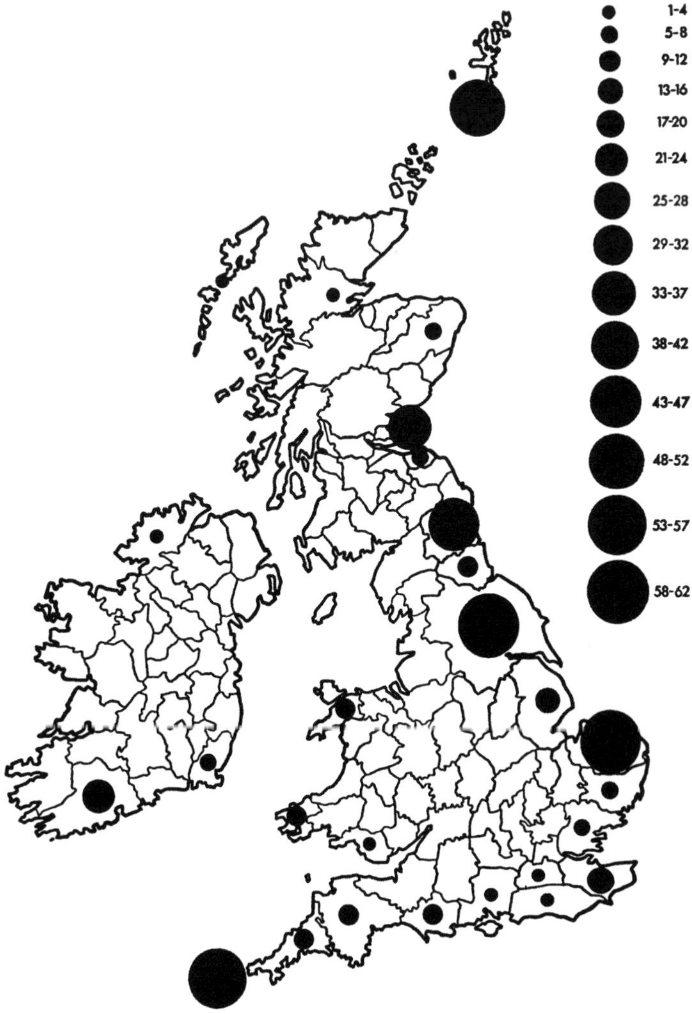

Fig. 127. Distribution by counties of autumn Red-breasted Flycatchers *Ficedula parva* in Britain and Ireland during 1958–67

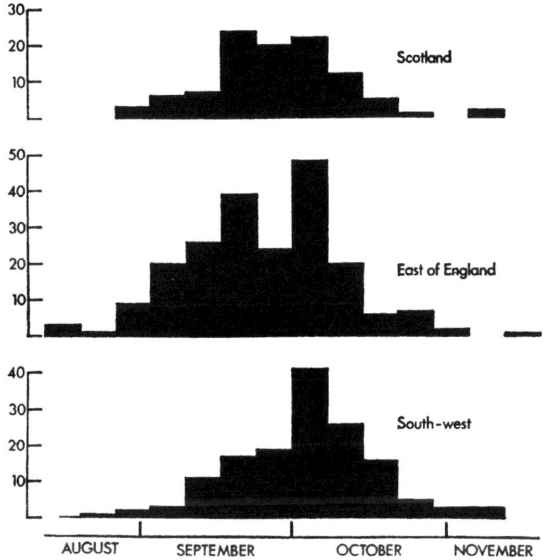

Fig. 128. Seasonal pattern of autumn Red-breasted Flycatchers *Ficedula parva* during 1958–67 with the records in Scotland, the east of England (Northumberland to Suffolk) and the south-west (south-west England, Wales and the south of Ireland) shown separately

between 48°N and 66°N; there is also a separate population lying east of the Black Sea and south of the Caspian (Vaurie 1959, Voous 1960). The eastern race *F. p. albicilla* (breeding east of Perm and the southern Urals) has never been recorded in Britain and Ireland, but few have been racially identified and this subspecies may occur (British Ornithologists' Union 1971). Grote (1940) stated that the whole population migrates south-east in autumn, but more recent evidence suggests that the western birds migrate east through Iran and Afghanistan to winter mainly in western India (Vaurie 1959) and occasionally perhaps in tropical east Africa (Voous 1960). The breeding range has been expanding westwards (Niethammer 1951), but this expansion seems to have been fairly slow: the first proved case of breeding in the Netherlands was not until 1967, despite several instances of possible nesting since the middle of the last century (Tekke 1967), and the fairly recent populations in Scandinavia are still small.

The discussions regarding the vagrancy pattern of Red-breasted Flycatchers have followed much the same sequence as those on Barred

Fig. 129. European distribution of Red-breasted Flycatchers *Ficedula parva* with the breeding range of this summer visitor shown in black (reproduced, by permission, from the 1966 edition of the *Field Guide*)

Warblers (see page 167). Rudebeck (1956), Vielliard (1962) and Diesselhorst (1971) have suggested that a small proportion migrates south-west in autumn, leading to a migratory divide reminiscent of a number of species, such as the Blackcap *S. atricapilla* (Williamson 1963c). Williamson (1959c, 1962c, 1963a) suggested that there was a random juvenile dispersal, perhaps especially in years of high population, with arrival in Britain caused by down-wind drift. Lack (1960) agreed with juvenile dispersal but not with down-wind drift, and left the question unsolved. Nisbet (1962), Rabøl (1969b) and Gatter (1972) have concluded that reverse migration, especially of juveniles, under the influence of central European anticyclones, explains the occurrences west of the breeding range. Radford (1968) has already analysed the autumn British and Irish records and concluded that the only possibilities were unoriented dispersal or a south-westward migration (Williamson's or Rudebeck's suggestions), though she ignored the hypothesis put forward by Nisbet (1962) and later by Rabøl and Gatter. As pointed out by Gatter, Diesselhorst took no account of the numerous records north and west of the breeding range and considered only the few known south European and north African occurrences when supporting Rudebeck's suggestion of a small south-westward migration. The scarcity of records from south-west Europe in comparison with those from north-west Europe has been taken as evidence against Rudebeck's idea, though this is clearly partly due to a shortage of observers. Isenmann (1971) reported twelve records from the Camargue during 1956–70 (one in April and the rest during September–December, with seven in October); and Gatter listed only four for Spain, but to these can be added one more (Wallace and Sage 1969).

Departure from eastern Germany starts in late July and the last birds have left by mid-September (Weber 1965): this links well with peaks on Christians Ø, off Bornholm, Denmark, in early September and at Blavand, Denmark, in late September (Rabøl 1969b), with 73 in the Netherlands since 1953 reaching a peak in early October (Gatter 1972), and with those in Britain and Ireland (Fig. 124).

There is clearly some connection between occurrences of Barred Warblers and Red-breasted Flycatchers, for in Britain and Ireland both had peak years in 1959 and 1965: this connection may have been related to the development of anticyclonic conditions initiating reverse migration, or it may simply have reflected good breeding success for both species in those years. Conversely, in 1963 Barred Warblers were numerous but Red-breasted Flycatchers fewer than in any other of the ten years. Whatever the influence, however, it clearly affects Barred Warblers more than Red-breasted Flycatchers, for the latter showed little annual variation in numbers, whereas the former fluctuated violently from year to year.

The spring birds (mainly in May–June) are too few for detailed analysis. Arrival in Germany is from mid-May onwards (Weber 1965). Gatter (1972) noted that there have been April records in Cyprus, Italy, southern France and Spain, 31 spring records on Heligoland (against 68 in autumn) with a peak in late May, and a few in the Netherlands, mainly in the second week of June. Rabøl (1969b) showed that spring records reached a peak in mid-May to early June at Christians Ø, and that they exceeded autumn records. Although inconclusive, the timing and distribution of the British and Irish records favour overshooting rather than a return on a SW–NE route, though the latter cannot be ruled out.

The geographical distribution of autumn records in Britain and Ireland alone does not immediately suggest a mirroring of the breeding range due to reverse migration as in the Barred Warbler, but, since the normal wintering area of Red-breasted Flycatchers is some 3,000 km farther east, and the standard direction is therefore likely to be ESE, this is hardly surprising. The pattern (Fig. 127) is far closer to those of the Greenish Warbler (Fig. 81) and Yellow-browed Warbler (Fig. 107), both of which winter in Asia and have an easterly standard direction, than to those of the Barred Warbler (Fig. 121) and Scarlet Rosefinch (Fig. 87), which winter farther to the west and have a south-easterly standard direction. The proportion of Red-breasted Flycatchers occurring in Scotland is intermediate between those of each of these pairs, as is the standard direction. The arrivals on the English east coast, particularly Yorkshire and Norfolk, earlier and in larger numbers than those in Scotland also accord with reverse migration as the cause. The data do not discount the possibility that western birds are derived from earlier arrivals in eastern Britain (unlike Barred Warblers, for example), but a direct south-eastern origin from the Continent seems more likely.

As concluded by Nisbet (1962), Rabøl (1969b) and Gatter (1972), reverse migration satisfactorily accounts for the patterns of occurences in western Europe in autumn and is probably the major cause of vagrancy in Britain and Ireland. The presence of adults in early autumn in Britain, and the admittedly small numbers in southern France, suggest that the tiny south-westward passage first proposed by Rudebeck (1956) may also occur.

CHAPTER 10

Summary

Spring, autumn and grand totals, and peak months are listed in Table 14, on p. 182. In 17 of the 24 species, and in Nearctic waders and landbirds as a whole, the numbers recorded in autumn were at least three times those in spring; in three (Temminck's Stint, Mediterranean Gull and White-winged Black Tern) the numbers in autumn were about twice those in spring; in two (Gull-billed Tern and Woodchat Shrike) the spring and autumn numbers were roughly equal; and only in Hoopoe and, especially, Golden Oriole did the spring numbers greatly exceed those in autumn. Table 14 also shows the peak springs and autumns for each species during the ten years.

The distribution maps are all brought together in a summarised form (by the twelve regions) in Fig. 130. These maps show five main distribution patterns. The following lists summarise the main areas of Britain and Ireland where each species was recorded during its main passage periods. Species included under more than one heading at a particular season are shown in italics; S = spring, A = autumn.

Northern east coast (Scotland and north-east England):

Long-tailed Skua (S)
Bluethroat (S)
Icterine Warbler (S)
Barred Warbler
Arctic Warbler (A)
Yellow-browed Warbler (A)
Red-breasted Flycatcher
Richard's Pipit (A)
Scarlet Rosefinch
Ortolan Bunting (S)

Southern east coast (eastern England and East Anglia):

Long-tailed Skua (A)
Icterine Warbler (A)
Red-breasted Flycatcher (A)
Woodchat Shrike (S)

South-east (East Anglia and south-east England):

Rough-legged Buzzard
Temminck's Stint
Pectoral Sandpiper (A)
Mediterranean Gull
White-winged Black Tern
Gull-billed Tern
Bluethroat (A)
Tawny Pipit

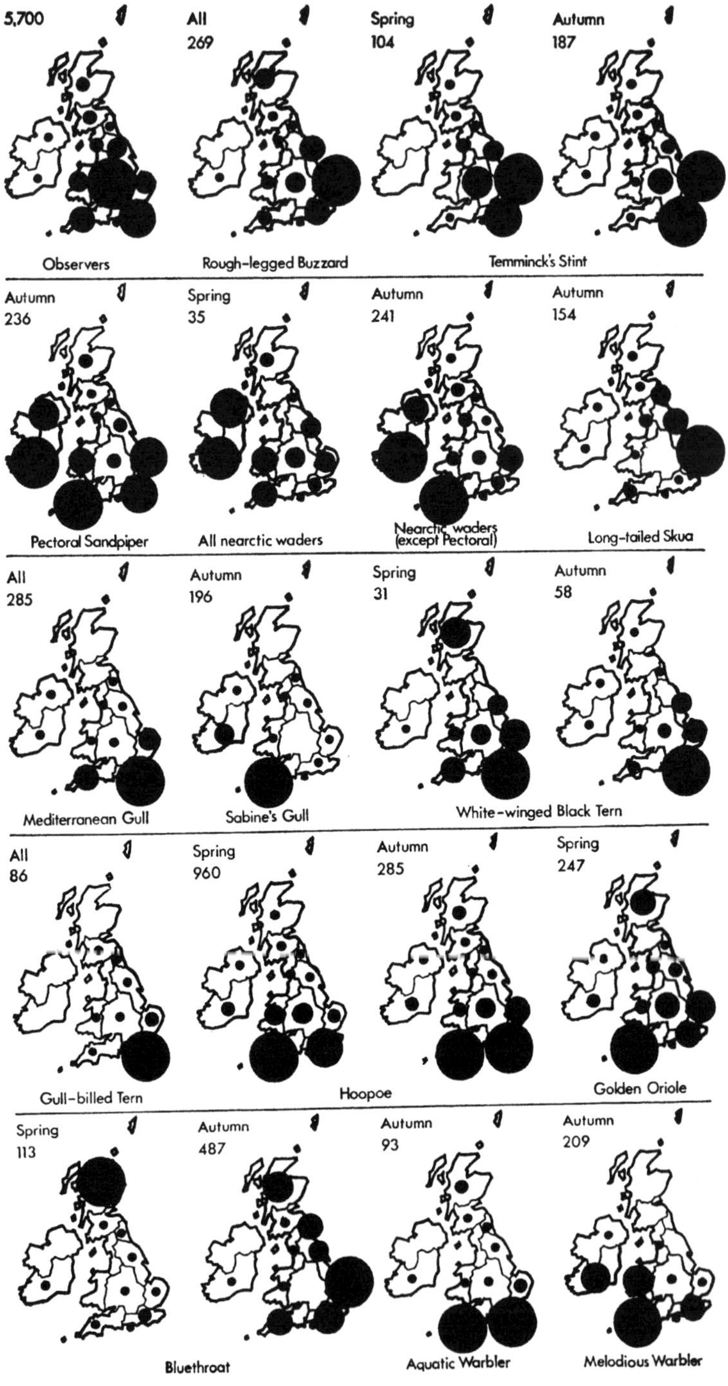
5,700
Observers
All
269
Rough-legged Buzzard
Spring
104
Autumn
187
Temminck's Stint
Autumn
236
Pectoral Sandpiper
Spring
35
All nearctic waders
Autumn
241
Nearctic waders
(except Pectoral)
Autumn
154
Long-tailed Skua
All
285
Mediterranean Gull
Autumn
196
Sabine's Gull
Spring
31
Autumn
58
White-winged Black Tern
All
86
Gull-billed Tern
Spring
960
Autumn
285
Hoopoe
Spring
247
Golden Oriole
Spring
113
Autumn
487
Bluethroat
Autumn
93
Aquatic Warbler
Autumn
209
Melodious Warbler

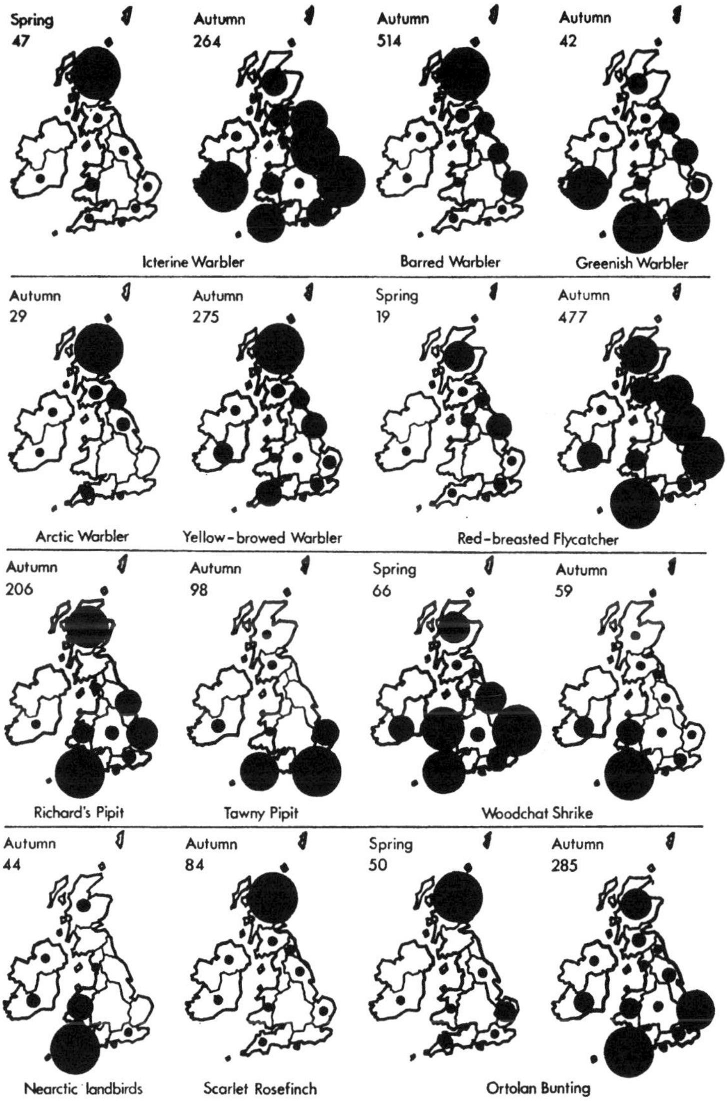

Fig. 130. Summary maps of distribution by regions of scarce migrants in Britain and Ireland during 1958–67. The season (all records, spring or autumn) is shown at the top of each map, where the total number of individuals recorded at that season in the ten years is also shown. (Some totals differ from those in chapters 1–9 for reasons indicated in the Introduction.) Eight sizes of dot are used, the largest in the region recording the highest number of birds and the others representing equal divisions of this (for example, if the largest dot represents 160 birds, the third size up is used for 41–60 birds). The extreme top left map shows the distribution of observers (taken from Fig. 1 in the Introduction), and the bias introduced by this—especially the few observers in Scotland and the even smaller number in Ireland—should be constantly borne in mind

Table 14. Seasonal and grand totals, peak months and peak seasons for 24 scarce migrants and for Nearctic waders and landbirds in Britain and Ireland during 1958–67

Under 'Totals' and 'Peak seasons', S = spring, A = autumn; capitals indicate main peaks, small letters secondary ones. Totals of the three species with asterisks cannot easily be split into spring and autumn

	TOTALS				Peak seasons									
	S	A	Grand	Peak months	58	59	60	61	62	63	64	65	66	67
Rough-legged Buzzard *Buteo lagopus*	*	*	269	October–November	—	—	A	—	a	—	—	—	A	—
Temminck's Stint *Calidris temminckii*	104	187	291	May, August–September	—	—	—	SA	—	A	S	A	S	S
Pectoral Sandpiper *Calidris melanotos*	7	236	234	September	—	—	—	A	A	—	—	—	—	A
Other Nearctic waders	28	241	269	September–October	—	—	—	—	—	A	—	—	A	A
Long-tailed Skua *Stercorarius longicaudus*	17	154	171	August–October	—	A	—	A	S	A	—	—	—	S
Mediterranean Gull *Larus melanocephalus*	*	*	285	July–August	—	—	—	A	—	—	A	A	A	A
Sabine's Gull *Larus sabini*	7	196	203	September–October	—	—	a	—	—	—	—	—	—	A
White-winged Black Tern *Chlidonias leucopterus*	31	58	89	May, August	—	s	S	S	—	—	sA	—	—	A
Gull-billed Tern *Gelochelidon nilotica*	*	*	86	Apr–May, Jun–Jul, Aug–Sept	—	—	SA	—	—	—	—	—	—	A
Hoopoe *Upupa epops*	960	285	1,245	April–May	SA	—	S	—	—	—	S	S	—	—
Golden Oriole *Oriolus oriolus*	247	10	257	May	—	—	—	—	—	—	S	S	—	S
Bluethroat *Luscinia svecica*	113	487	600	September	SA	S	A	—	—	—	—	A	—	—
Aquatic Warbler *Acrocephalus paludicola*	1	93	94	August–September	a	A	—	a	—	—	—	—	—	—
Melodious Warbler *Hippolais polyglotta*	8	209	217	August–September	—	—	—	—	A	—	—	a	—	a
Icterine Warbler *Hippolais icterina*	47	264	311	August–September	—	—	—	—	—	—	—	a	A	S
Barred Warbler *Sylvia nisoria*	2	514	516	August–September	—	A	—	—	—	a	—	A	A	a
Greenish Warbler *Phylloscopus trochiloides*	4	42	46	September–October	—	—	—	A	A	—	A	—	—	A
Arctic Warbler *Phylloscopus borealis*	0	29	29	September	—	a	—	—	—	—	a	—	—	a
Yellow-browed Warbler *Phylloscopus inornatus*	0	275	275	September–October	—	—	a	—	—	—	a	—	—	A
Red-breasted Flycatcher *Ficedula parva*	19	477	496	September–October	—	A	—	—	—	—	—	A	a	a
Richard's Pipit *Anthus novaeseelandiae*	6	206	212	September–October	—	—	—	—	—	—	—	—	A	A
Tawny Pipit *Anthus campestris*	13	98	111	September	A	—	A	—	—	—	—	A	—	A
Woodchat Shrike *Lanius senator*	66	59	125	May–Jun, Aug–Sept	A	—	S	—	—	—	s	sA	—	SA
Nearctic landbirds	13	44	57	October	—	—	—	—	A	—	—	—	sA	A
Scarlet Rosefinch *Carpodacus erythrinus*	7	84	91	September	—	A	—	—	—	—	A	—	A	A
Ortolan Bunting *Emberiza hortulana*	50	285	335	September	S	—	A	A	—	—	—	A	A	S

South coast (south-east and south-west England):

Mediterranean Gull
Hoopoe
Aquatic Warbler (A)
Greenish Warbler (A)
Tawny Pipit

South-west (south-west England, Wales and the south of Ireland):

Pectoral Sandpiper
Other Nearctic waders
Sabine's Gull (A)
Hoopoe (S)
Golden Oriole (S)
Melodious Warbler
Icterine Warbler (A)
Greenish Warbler (A)
Red-breasted Flycatcher (A)
Richard's Pipit (A)
Woodchat Shrike
Nearctic landbirds (A)
Ortolan Bunting (A)

The reasons for the occurrence of these species in Britain and Ireland that have been tentatively suggested in the various chapters may be summarised in the following lists. When an explanation is based on reasonable evidence, the species is listed in ordinary type; when a pure guess is involved, italics and a question mark are used. This categorisation undoubtedly oversimplifies the situation (even if the positioning within the categories is correct), and reference should be made to the individual species accounts.

Overshooting in anticyclonic conditions in spring:

White-winged Black Tern?
Hoopoe
Golden Oriole
White-spotted Bluethroat
Melodious Warbler?
Icterine Warbler?
Greenish Warbler?
Red-breasted Flycatcher
Tawny Pipit
Woodchat Shrike
Scarlet Rosefinch
Ortolan Bunting?

Britain and Ireland within usual migration path (number of occurrences dependent upon weather to the extent of shifting the route slightly from year to year):

Rough-legged Buzzard (larger numbers in invasion years)
Temminck's Stint
Long-tailed Skua
Mediterranean Gull
Sabine's Gull
Gull-billed Tern
Red-spotted Bluethroat (A)
Ortolan Bunting (A, east coast)

Usual migration path near to Britain and Ireland (weather causing deviation to varying extent each year):

White-winged Black Tern (s)?
Hoopoe (A)
Red-spotted Bluethroat (s)
White-spotted Bluethroat (A)
Aquatic Warbler (A)
Icterine Warbler (s, and east coast A)
Tawny Pipit (A)
Ortolan Bunting (s)?

Reverse migration in autumn (supposed orientations as follows).

N	*Melodious Warbler?*
NNW	*Icterine Warbler?*
	Woodchat Shrike?
NW	Barred Warbler
	Scarlet Rosefinch
WNW	Red-breasted Flycatcher
W	Greenish Warbler
	Arctic Warbler
	Yellow-browed Warbler
	Richard's Pipit?

Random post-breeding dispersal:

White-winged Black Tern?
Melodious Warbler?
Icterine Warbler?
Richard's Pipit?
Woodchat Shrike?
Ortolan Bunting?

Transatlantic crossings initiated and/or aided by westerly gales:

Pectoral Sandpiper
Other Nearctic waders
Nearctic landbirds

References

To save space, in view of the large number of references cited in this book, the following list is given in abbreviated form, with titles of papers and articles omitted and authors' surnames grouped in paragraphs according to their initial letters.

Aikman, E. F. (1966): *Sea Swallow*, 18: 51–53, 62–64. **Alexander, W. B., and Fitter, R. S. R.** (1955): *Brit. Birds*, 48: 1–14. **Anon.** (1949): *Brit. Birds*, 42: 135–143; (1951a): *Brit. Birds*, 44: 250–252; (1951b): *Brit. Birds*, 44: 254–256, 420–421; (1952a): *Brit. Birds*, 45: 18; (1952b): *Brit. Birds*, 45: 294; (1952c): *Brit. Birds*, 45: 416; (1957): *Brit. Birds*, 50: 441. **Aumees, L., and Paakspuu, V.** (1963): *Orn. Kogumik*, 3: 195–205.

Baird, J., Bagg, A. M., Nisbet, I. C. T., and Robbins, C. S. (1959): *Bird-Banding*, 30: 143–171. **Bannerman, D. A.** (1953–62): *The Birds of the British Isles* (Edinburgh and London). **Barnes, J. A. G.** (1960): *Brit. Birds*, 53: 88–89. **Baxter, E. V., and Rintoul, L. J.** (1953): *The Birds of Scotland* (Edinburgh and London). **Bell, D. G.** (1965): *Brit. Birds*, 58: 139–145. **Beretzk, P.** (1955): *Aquila*, 62: 430–431. **Blondel, J.** (1966): *Terre et vie*, 35: 237–254. **Bonham, P. F.** (1970): *Brit. Birds*, 63: 145–147; (1971): *Brit. Birds*, 64: 135–136. **Bonham, P. F., and Bonham, M. R.** (1971); *Brit. Birds*, 64: 409–411. **Borgstrom, E.** (1967): *Vår Fågelv.*, 26: 54–56. **Bourne, W. R. P.** (1965): *Seabird Bull.*, 1: 34–37; (1967): *Sea Swallow*, 19: 51–76; (1970a): *Sea Swallow*, 20: 53; (1970b): *Brit. Birds*, 63: 91–93. **British Ornithologists' Union** (1971): *The Status of Birds in Britain and Ireland* (Oxford, London and Edinburgh). **Braun, B.** (1971): *Brit. Birds*, 64: 385–408.

Christensen, S., Neilsen, B. P., Porter, R. F., and Willis, I. (1971–73): *Brit. Birds*, 64: 247–266, 435–455; 65: 52–78, 233–247, 411–423; 66: 100–114, 285–298, 472–493. **Commissie voor de Nederlandse Avifauna** (1970): *Avifauna van Nederland* (Leiden). **Cramp, S.** (1968): *Brit. Birds*, 61: 405–408. **Curry-Lindahl, K.** (1961): *Brit. Birds*, 54: 297–306.

Dambiermont, J.-L., Fouarge, J., and Delvaux, L. (1964): *Aves*, 1: 77–84. **Davenport, L. J.** (1973): *Brit. Birds*, 66: 497–498. **Davis, P.** (1962): *Fair Isle Bird Obs. Bull.*, 4: 228–229; (1964a): *Bird Study*, 11: 198–223; (1964b): *Brit. Birds*, 62: 214–216; (1964c): *Fair Isle Bird Obs. Bull.*, 5: 115; (1966): *Brit. Birds*, 59: 353–376; (1967): *Bird Study*, 14: 65–95. **de Cock de Rameyen, D., and Flamand, V.** (1968): *Gerfaut*, 58: 148–151. **Dementiev, G. P., and Gladkov, N. A.** (1951–54): *The Birds of the Soviet Union* (Moscow). **Dennis, R. H.** (1967a): *Fair Isle Bird Obs. Bull.*, 5: 245–246; (1967b): *Fair Isle Bird Obs. Report for 1966*; (1968): *Fair Isle Bird Obs. Report for 1967*; (1969): *Fair Isle Bird Obs.*

Report for 1968; (1970): *Fair Isle Bird Obs. Report for 1969*, 11–23; (1972): *Brit. Birds*, 65: 481. **Devillers, P.** (1964): *Aves*, 1: 13–16. **Diesselhorst, G.** (1971): *Anz. Orn. Ges. Bayern*, 10: 187–188. **Dolnik, V. R., and Shumakov, M. E.** (1967): *Bionica*, 500–507. **Durand, A. L.** (1963): *Brit. Birds*, 56: 157–164; 1972: *Brit. Birds*, 65: 428–442.

England, M. D. (1971): *Brit. Birds*, 64: 421. **Editors** (1949): *Brit. Birds*, 42: 89; 43: 224; (1951): *Brit. Birds*, 44: 204–205, 419–420. **Evans, P. R.** (1968): *Brit. Birds*, 61: 281–303.

Feeny, P. P. (1959): *Bird Migration*, 1: 153–158. **Fenton, J. K., and Cudworth, J.** (1968): *Spurn Bird Obs. Report for 1967*. **Ferguson-Lees, I. J.** (1952): *Brit. Birds*, 45: 357–358; (1954): *Brit. Birds*, 47: 121–123; (1955a): *Brit. Birds*, 48: 75–76; (1955b): *Brit. Birds*, 48: 356–357; (1955c): *Brit. Birds*, 48: 499–500; (1958): *Brit. Birds*, 51: 205; (1963): *Brit. Birds*, 56: 140–148; (1964): *Brit. Birds*, 57: 337–340; (1965a): *Brit. Birds*, 58: 9–10; (1965b): *Brit. Birds*, 58: 461–464; (1967): *Brit. Birds*, 60: 345–347. **Ferguson-Lees, I. J., and Sharrock, J. T. R.** (1967): *Brit. Birds*, 60: 534–540. **Ferguson-Lees, I. J., and Williamson, K.** (1960): *Brit. Birds*, 53: 529–544; (1961): *Brit. Birds*, 54: 44–48. **Festetics, A.** (1959): *Egretta*, 4: 67–74. **Fisher, J., and Lockley, R. M.** (1953): *Sea-Birds* (London). **Fjerdingstad, C.** (1939): *Alauda*, 11: 50–54.

Gatter, W. (1972): *Die Vogelwelt*, 93: 91–98. **Glegg, W. E.** (1942): *Ibis*, 84: 390–434. **Goodwin, D.** (1956): *Brit. Birds*, 49: 339–349. **Grant, P. J.** (1972): *Brit. Birds*, 65: 287–290. **Grant, P. J., and Scott, R. E.** (1967): *Brit. Birds*, 60: 365–368. **Greenwood, J. J. D.** (1968): *Brit. Birds*, 61: 524–525. **Grote, H.** (1940): *J. Orn.*, 88: 355–372.

Harber, D. D. (1955): *Brit. Birds*, 48: 313–319; (1964a): *Brit. Birds*, 57: 211–213; (1964b); *Brit. Birds*, 57: 261–281; (1966): *Brit. Birds*, 59: 280–305. **Headlam, C. G.** (1972): *Scot. Birds*, 7: 94. **Hollom, P. A. D.** (1957): *Brit. Birds*, 50: 73–75. **Hudson, R.** (ed.) (1971): *A Species List of British and Irish Birds* (Tring). **Hume, R. A., and Lansdown, P. G.** (1974): *Brit. Birds*, 67: 17–24.

Isenmann, P. (1971): *Anz. Orn. Ges. Bayern*, 10: 187–188.

Japin, H. J., and Van der Velden, B. (1959): *Limosa*, 32: 183–185. **Jozefik, M.** (1960): *Acta Orn. Mus. Zool. Polon.*, 5: 307–324.

Kennedy, P. G., Ruttledge, R. F., and Scroope, C. F. (1954): *Birds of Ireland* (Edinburgh and London).

Lack, D. (1960): *Brit. Birds*, 53: 325–352, 379–397; (1961): *Bird Migration*, 2: 49–51. **Lippens, L.** (1968): *Gerfaut*, 58: 163; (1970): *Gerfaut*, 60: 26–40. **Liversidge, R., and Courtenay-Latimer, M.** (1963): *Ann. Cape Prov. Mus.*, 3: 57–60. **Lundberg, S., Hojer, J., and Norbeck, J.** (1954): *Vår Fågelv.*, 13: 244.

Mayaud, N. (1954): *Alauda*, 22: 225–245; (1956): *Alauda*, 24: 123–131; (1961): *Alauda*, 29: 165–173; (1965): *Alauda*, 33: 81–83. **Merikallio, E.** (1958):

Fauna Fenn., 5: 1–181. **Mikelsvaar, N.** (1963): *Orn. Kogumik*, 3: 148–158. **Morgan, K., and Wheeler, P.** (1958): *Ostrich*, 39: 90. **Muller, A. K.** (1967): *Vogelwarte*, 24: 63–64.

Nelson, T. H. (1907): *The Birds of Yorkshire* (London). **Nicholson, E. M., and Ferguson-Lees, I. J.** (1962): *Brit. Birds*, 55: 299–384. **Niethammer, G.** (1937): *Handbuch der Deutschen Vogelkunde* (Leipzig); (1951): *Bonn Zool. Beitr.*, 2: 17–54. **Nisbet, I. C. T.** (1957): *Brit. Birds*, 50: 197–200; (1959): *Brit. Birds*, 52: 205–215; (1962): *Brit. Birds*, 55: 74–86; (1963): *Brit. Birds*, 56: 204–217. **Norregaard, K.** (1964): *Dansk Orn. Foren Tidsskr.*, 58: 42–43.

Palm, B. (1952): *Dansk Orn. Foren. Tidsskr.*, 46: 32–52. **Parslow, J. L. F.** (1967–68): *Brit. Birds*, 60: 2–47, 97–123, 177–202, 261–285, 396–404, 493–508; 61: 49–64, 241–255; (1973): *Breeding Birds of Britain and Ireland* (Berkhamsted). **Peterson, R. T.** (1947): *A Field Guide to the Birds.* (Boston, Mass.). **Peterson, R., Mountfort, G., and Hollom, P. A. D.** (1954, 1966): *A Field Guide to the Birds of Britain and Europe* (London) (referred to in this book as 'the *Field Guide*'). **Pettitt, R. G.** (1969): *Seabird Bull.*, 7: 10–23. **Phillips, N. R.** (1966): *Seabird Bull.*, 2: 11–15. **Preuss, N.O.** (1966): *Feltornithologen*, 8: 1–22. **Pyman, G. A., and Wainwright, C. B.** (1952): *Brit. Birds*, 45: 337–339.

Rabøl, J. (1969a): *Brit. Birds*, 62: 89–92; (1969b): *Feltornithologen*, 11: 123–131; (1970): *Feltornithologen*, 12: 87–89. **Radford, M. C.** (1968): *Bird Study*, 15: 154–160. **Ricard, M.** (1966): *Oiseau*, 36: 64. **Richardson, R. A.** (1962): *Check-list of the Birds of Cley and Neighbouring Norfolk Parishes.* **Robbins, C. S., Brunn, B., and Zim, H. S.** (1966): *Birds of North America.* (Racine, Wisconsin). **Rogers, M. J.** (1972): *Brit. Birds*, 65: 37. **Rosin, K., and Wagner, S.** (1964): *J. Orn.*, 105: 85–86. **Roux, F.** (1960): *Nos Oiseaux*, 25: 315–317; (1961): *Alauda*, 29:161–164. **Rudebeck, G.** (1956): *Hanström Festschrift*, 257–268. **Ruttledge, R. F.** (1966): *Ireland's Birds* (London).

Sage, B. L. (1968): *Ibis*, 110: 1–14. **Schevareva, T. P.** (1955): *Trudy Biuro Kolitsevaniya*, 8: 46–90. **Scott, R. E.** (1968): *Brit. Birds*, 61: 449–455; (1969): *Brit. Birds*, 62: 84–85. **Sharrock, J. T. R.** (1965a): *Brit. Birds*, 58: 520–521; (1965b): *Cape Clear Bird Obs. Rep.*, 6: 50–56; (1966): *Brit. Birds*, 59: 556–558; (1967): *Seabird Bull.*, 3: 21–26; (1968a): *Bird Study*, 15: 99–103; (1968b): *Bird Study*, 15: 214; (1969–73): *Brit. Birds*, 62: 169–189, 300–315; 63: 6–23, 313–324; 64: 93–113, 302–309; 65: 187–202, 381–392; 66: 46–64, 517–525; (1973): *The Natural History of Cape Clear Island* (Berkhamsted). **Smith, F. R.** (1967): *Brit. Birds*, 60: 309–338; (1968): *Brit. Birds*, 61: 329–365; (1969): *Brit. Birds*, 62: 457–492; (1970): *Brit. Birds*, 63: 267–293. **Sorensen, L. H., Kramshoj, E., and Christensen, N. H.** (1964): *Brit. Birds*, 57: 213–214. **Spencer, R.** (1957): *Brit. Birds*, 50: 449–485; (1959): *Brit. Birds*, 52: 441–449; (1960): *Brit. Birds*, 53: 457–502; (1963): *Brit. Birds*, 56: 477–524; (1964): *Brit. Birds*, 57: 525–582; (1965): *Brit. Birds*, 58: 533–583; (1968): *Brit. Birds*, 61: 477–523; (1969): *Brit. Birds*, 62: 393–442. **Svensson, L.**, (1970): *Identification Guide to European Passerines* (Stockholm).

Taverner, J. H. (1970): *Brit. Birds*, 63: 67–79; (1972): *Brit. Birds*, 65: 185–186. **Tekke, M. J.** (1967): *Limosa*, 40: 186–187; (1968): *Limosa*, 41: 31–34; (1973): *Brit. Birds*, 66: 540–541. **Thomson, A. L.** (1948): *Brit. Birds*, 41: 295–300. **Thomson, A. L., and Leach, E. P.** (1953): *Brit. Birds*, 46: 313–330. **Tuck, G. S.** (1966): *Sea Swallow*, 18: 47–50, 54; (1970): *Sea Swallow*, 20: 36–38. **Turcek, F. J.** (1962–63): *Biologica*, 16: 511–523; 18: 504–514.

Valikangas, I. (1951a): *Proc. Int. Orn. Congr.*, 10: 527–531; (1951b): *Ornis Fennica*, 28: 25–39. **Valverde, J. A.** (1968): *Ardeola*, 12: 117–120. **Vaurie, C.** (1959, 1965): *The Birds of the Palearctic fauna* (London). **Veroman, H.** (1963): *Orn. Kogumik*, 3: 159–175. **Vielliard, J.** (1962): *Oiseau*, 32: 74–79. **Voous, K. H.** (1960): *Atlas of European Birds* (Amsterdam and London).

Wallace, D. I. M. (1964): *Brit. Birds*, 57: 282–301; (1970): *Brit. Birds*, 63: 113–129. **Wallace, D. I. M., and Grant, P. J.** (1974): *Brit. Birds*, 67: 1–16. **Wallace, D. I. M. and Sage, B. L.** (1969): *Ardeola*, 14: 143–157. **Weber, H.** (1965): *Brit. Birds*, 58: 434–438. **Wehner, R.** (1966): *Vogelwarte*, 23: 173–180; (1967): *Vogelwarte*, 24: 64. **Williamson, K.** (1958): *Bird Migration*, 1: 27–36; (1959a): *Brit. Birds*, 52: 334–377; (1959b): *Bird Migration*, 1: 84–88; (1959c): *Bird Migration*, 1: 147–152; (1960a): *Brit. Birds*, 53: 243–252; (1960b): *Bird Migration*, 1: 161–170; (1960c): *Identification for Ringers. 1. The Genera Locustella, Lusciniola, Acrocephalus and Hippolais*; (1961a): *Bird Migration*, 2: 1–33; (1961b): *Bird Migration*, 2: 51–53; (1961c): *Bird Migration*, 1: 218–234; (1962a): *Identification for Ringers. 2. The Genus Phylloscopus*; (1962b): *Brit. Birds*, 55: 130–131; (1962c): *Bird Migration*, 2: 61–102; (1962d): *Bird Migration*, 2: 131–159; (1963a): *Bird Migration*, 2: 207–223; (1963b): *Bird Migration*, 2: 224–251; (1963c): *Bird Migration*, 2: 265–271; (1963d): *Brit. Birds*, 56: 285–292; (1964): *Identification for Ringers. 3. The Genus Sylvia*. **Williamson, K., and Ferguson-Lees, I. J.** (1960): *Brit. Birds*, 53: 369–378. **Witherby, H. F., Jourdain, F. C. R., Ticehurst, N. F., and Tucker, B. W.** (1938–41): *The Handbook of British Birds* (London) (referred to in this book as *The Handbook*). **Wittgen, A. B., and Braaksma, S.** (1964): *Limosa*, 37: 12–15.

Zoutendyk, P. (1965): *Ostrich*, 36: 15–16.

Index

The page numbers of the main species accounts are shown in *italics*.